Nanotechnology in the Life Sciences

Series Editor

Ram Prasad
Department of Botany
Mahatma Gandhi Central University
Motihari, Bihar, India

Nano and biotechnology are two of the 21st century's most promising technologies. Nanotechnology is demarcated as the design, development, and application of materials and devices whose least functional make up is on a nanometer scale (1 to 100 nm). Meanwhile, biotechnology deals with metabolic and other physiological developments of biological subjects including microorganisms. These microbial processes have opened up new opportunities to explore novel applications, for example, the biosynthesis of metal nanomaterials, with the implication that these two technologies (i.e., thus nanobiotechnology) can play a vital role in developing and executing many valuable tools in the study of life. Nanotechnology is very diverse, ranging from extensions of conventional device physics to completely new approaches based upon molecular self-assembly, from developing new materials with dimensions on the nanoscale, to investigating whether we can directly control matters on/in the atomic scale level. This idea entails its application to diverse fields of science such as plant biology, organic chemistry, agriculture, the food industry, and more.

Nanobiotechnology offers a wide range of uses in medicine, agriculture, and the environment. Many diseases that do not have cures today may be cured by nanotechnology in the future. Use of nanotechnology in medical therapeutics needs adequate evaluation of its risk and safety factors. Scientists who are against the use of nanotechnology also agree that advancement in nanotechnology should continue because this field promises great benefits, but testing should be carried out to ensure its safety in people. It is possible that nanomedicine in the future will play a crucial role in the treatment of human and plant diseases, and also in the enhancement of normal human physiology and plant systems, respectively. If everything proceeds as expected, nanobiotechnology will, one day, become an inevitable part of our everyday life and will help save many lives.

Ayesha Nazeer • Faisal Ahmad
Neeraj Verma • Shamim Ahmad

Targeted Delivery of Nanopesticides and Nanofertilizers in Sustainable Agricultural Farming

Springer

Ayesha Nazeer
Amunix Pharmaceuticals Inc.
South San Francisco, CA, USA

Faisal Ahmad
Iris Worldwide
Gurugram, India

Neeraj Verma
Amicable Knowledge Solution University
Satna, Madhya Pradesh, India

Shamim Ahmad
Centre of Excellence in Nanotechnology
New Delhi, India

ISSN 2523-8027 ISSN 2523-8035 (electronic)
Nanotechnology in the Life Sciences
ISBN 978-3-031-41332-2 ISBN 978-3-031-41333-9 (eBook)
https://doi.org/10.1007/978-3-031-41333-9

This Springer imprint is published by the registered company Springer Nature Switzerland AG
The registered company address is: Gewerbestrasse 11, 6330 Cham, Switzerland

Paper in this product is recyclable.

Preface

The proposed book is meant for students of graduate and doctoral degrees in the agri-nanobiotechnology discipline, where the main emphasis has been given to understanding the interactions of nanoscale materials and species of pesticides and fertilizers during their applications in the farm field to improve quality and increase yield with the minimum use of the active ingredients attached to the nanocarriers. This multidisciplinary approach involves expertise from fields like material science, nanotechnology, nanobiosensors, predictive mathematical modelling of large data derived from field trials, searching for appropriate marketing strategies, field trials of nanoformulations using *in vitro* and *in vivo*, and field trial experimentation, besides the core knowledge of the plant systems and biosphere. The authors, loaded with their respective specialized experiences, have tried to collect the appropriate details of the basic interactions, nanoscale synthesis, and application protocols to make them easy to appreciate and comprehend by the targeted readers. Giving sufficient hints in the text of the book at various places in relation to the less explored areas would be another plus point of this book for the aspiring graduate as well as postgraduate students to go for their research and development carrier based on these observations compiled from various latest reviews and research publications, which are given in the form of suggested readings. The text has been made to have a minimum number of references to make the reader concentrate more on the subject matter without much diversion.

We hope this approach to explaining the future perspectives in the highly relevant area of agri-nanobiotechnology will be relevant to pursue for innovative research and development of sustainable agriculture in the years to come.

South San Francisco, CA, USA — Ayesha Nazeer
Gurugram, Haryana, India — Faisal Ahmad
Satna, Madhya Pradesh, India — Neeraj Verma
New Delhi, India — Shamim Ahmad

Contents

About the Authors

Ayesha Nazeer, is a scientist with 20+ years of experience in the biopharmaceutical industry. With a background in Cell Biology, she has worked on the validation of novel drug targets in the drug discovery process. She started her career at Medarex, Inc., USA, which was later acquired by Bristol-Myers Squibb. After spending 17 years there in the Preclinical Pharmacology Group, she is currently working at Amunix Pharmaceuticals Inc., a startup based in South San Francisco. At Amunix, her work focuses on determining the safety profile of oncology targets.

Ayesha has a Master's degree in Biotechnology. She is settled in San Jose, California.

Faisal Ahmad, having more than 18 years of experience in turning marketing ideas into reality, has been responsible for steering the growth of Iris in Delhi by setting up its digital offering for clients 'GLOCALLY'. His expertise in developing solutions in the area of digital marketing has been quite successful with several industry brands like Airtel Africa, General Motors, MINI Cooper, Nokia, and Samsung, to name a few, in addition to having much wider coverage of valuable experience across industries like Telecom, IT, Automobile, Hospitality, BFSI, and Retail. Holding a Master of Engineering in Computer Science from BITS, Pilani, he spent the first 5 years of his experience as a Software Engineer before moving into managing larger digital mandates involving marketing technology aspects of the business. His current focus areas include 'Design Thinking for Integrated Marketing Offerings' and developing business solutions based on Internet of Things (IoT) concepts. Combining the recent developments in IoT with social IoT networks already in vogue, one can find out the options for solutions to avoid the emerging security issues. In this context, the networking aspects of the objects with low-cost printing technology for sensors and devices are also being studied by him and his research team from an industrial angle.

Neeraj Verma, Ph.D., with specialization in Botany (Plant Pathology), is working as Professor at the Department of Agriculture Science, Faculty of Agriculture Science and Technology, AKS University, Satna, India, since 2012. Earlier, he was Assistant Specialist (Plant Virology) at the University of California Riverside, Riverside, CA, USA. He has research and teaching experience spanning more than 20 years. His research areas include molecular plant virology, biotechnology, and nanobiotechnology. He has published more than 45 international and national research papers and book chapters and has published three books. He has worked on various research projects funded by government agencies. He has attended many national and international conferences. He is an editor and reviewer for several professional journals.

Shamim Ahmad started his R&D career at the Council of Scientific and Industrial Research (CSIR) – Central Electronics Engineering Research Institute (CEERI), Pilani, Rajasthan, India, and served there from 1972 to 2005. He contributed significantly not only to electronics for strategic applications but also actively participated in setting up R&D facilities and forming research groups to work in semiconductor physics and device applications before getting superannuated as Director. Later, he joined Hamdard University, New Delhi, as Vice Chancellor and succeeded in infusing R&D culture into the University. The University of Malaysia, Perlis, Malaysia, invited him as a Visiting Professor at the Institute of Nanoelectronics Engineering. After returning from Malaysia, he joined the Confederation of Indian Industry (CII) as Chief Nanotechnology Adviser to establish a Centre of Excellence in Green Nanotechnology in Ahmedabad, Gujrat, India, especially working on green nanotechnology processes. Thereafter, he joined the JC Bose University of Science and Technology YMCA, Faridabad, as Visiting Professor, at the Department of Electronics Engineering, and established a modular clean room facility at a reduced cost for academic institutions.

Chapter 1
Nanoscience, Nanotechnology and Engineered Nanomaterials: An Introduction

1.1 Introduction

With the ongoing development of "nanoscience" and "nanotechnology", a number of newer terms have very often been coined in research publications, giving identification to a new area of development using some specific class of nanoscale materials that are put to use. Apparently, there seems to be a situation that represents a complex mix of newer phenomena; however, once the meaning of the "nano" prefix is understood, most of these ensuing definitions become easy to understand and use appropriately.

There is another reason for this mixed state of affairs due to the fact that the experts reporting their scientific research often take the liberty of extending or shortening the well-known range of 1–100 nm, which is almost universally accepted in the scientific communities. However, the area they belong to also matters in defining this range according to their convenience. At times, considering a few hundred nm of dimension might appear justified. On the other hand, it may be confined to a few nm to explain a set of experimental results they are investigating. These ambiguities can be put to rest once it is assumed that the physicochemical properties of these nanoscale materials must be significantly different from their bulk counterparts. There is no hard and fast criteria to fix this limit beyond which it approaches the bulk material characteristics or vice versa in all the cases. This criterion for fixing an upper limit thus depends upon the material as well as its morphological structures.

Once the concept of nanotechnology is extended to biological entities, the terminology changes to "nanobiotechnology" or "bio nanotechnology," depending upon the discipline in which the research work has been carried out. Similarly, when it is related to agriculture, it acquires a term like "nano agriculture" and so on. These are the derivatives of nanotechnology extended to numerous fields of research and development.

A. Nazeer et al., *Targeted Delivery of Nanopesticides and Nanofertilizers in Sustainable Agricultural Farming*, Nanotechnology in the Life Sciences,
https://doi.org/10.1007/978-3-031-41333-9_1

Characteristic features of nanoscale materials arise due to changes in the arrangements of constituent atoms and molecules, giving them altogether different properties. In other words, inequilibrium structures at the nanoscale become distinctly different from those at thermodynamic equilibrium in bulk form. Chemical bonds formed with distortions become the reason for exhibiting physicochemical characteristics. Not only that, the abundance of unsaturated bonds present on the surface of such nanoscale materials makes the nanoscale surfaces highly chemically reactive and ever ready to establish some sort of bond with materials present in the environmental surroundings. This typical situation gives rise to fast aggregations among the nanoscale particulate species that change their properties altogether from what they were earlier in their synthesized state. Another typical characteristic of these highly active surfaces of material species present in an electrolyte is that they start interacting with charged molecular or ionic material species, resulting in a screening like situation in which the initial nanoscale material species act like a core surrounded by protein plasma, quite frequently encountered in biomolecular systems. Encouraged by this observation, it becomes possible to modify the surfaces by conjugating known ligand molecules to engineer the surfaces for specific functionalities. Consequently, the feasibility of attaching some molecular recognition motifs for not only guiding their interaction pathways but also realising targeted deliveries of specific molecular cargo belonging to pharmaceutical and molecular imaging activities.

Historically, nanoscale material species were more exhaustively investigated for their possible applications in biomedicine, including diagnostic imaging. Subsequently, with the advancements in the field of plant sciences based on the applications of microbiology, the concepts of nanoscale materials species started falling into the special domain of smart and intelligent agriculture. Using molecular species-based biosensors, it has now started imparting smart features into farming practices. Once biosensing is integrated with artificial intelligence, the possibilities of remote sensing and actuation could possibly combine to provide an intelligent solution to take care of farming problems in an integrated manner. The role of the Internet of Things (IoT) and related developments will provide the requisite support to implement intelligent solutions. These steps, once developed and implemented in the case of agricultural farming, would possibly be better able to take care of yield enhancement and sustainability by utilizing the existing resources in a reasonably optimal way.

In this chapter, effort has been made to present the standard definitions of the nanoscience and nanotechnology terms as recommended by various agencies with their massive programmes of development in the form of various initiatives and promoting related studies at national and global levels in the US and Europe, besides those groups joining from Asian regions. The contributions from China, Korea, and India are steadily increasing, as reflected in the increasing number of research publications and patents in the related areas year after year.

1.2 Definitions of Various Terms

1.2.1 Nanoscience

A generalized definition of nanoscience was given by the US-sponsored National Nanotechnology Initiative (NNI) programme as "the study and applications of extremely small things that might be used across all the other science fields, such as chemistry, biology, physics, materials science, and engineering," after going through an exhaustive exercise of evaluating the possible impacts of nanoscale materials anticipated in the future. Subsequent definitions included a more quantitative description in the form of "nanoscale dimensions," falling in the range of atomic size (0.2 nm) to 100 nm instead of "extremely small things," possibly based on the outcome of the ongoing R&D activities in the related fields. However, in this definition as well, the upper limit of the nanoscale, set at 100 nm, might still be considered tentative as there is no specific reason to exclude one above or below this upper limit. In the case of engineered nanomaterials, to be discussed later, the nanoscale dimension is limited to much less than 100 nm, i.e., 20–30 nm.

It may be clearly noted that the physicochemical properties of the nanoscale material species must be different from those reflected in their bulk counterparts. Thus, the lower limit remains the same as the atomic dimension, but the upper limit may be fixed, above which the material properties approach their bulk values. The basic concept that emerges from the above-mentioned definition seems to be dependent on the constituent atomic and molecular species confined in the nanoscale dimensions, imparting morphology specific physico-chemico-biological properties. The morphological changes introduced internally and externally, thus, offer the feasibility of altering the physico-chemico-biological characteristics of these extremely small material species falling in the nanoscience domain. This kind of manoeuvrability of physicochemical properties offers special leverage to discover newer materials with predetermined features, subsequently called "engineered nanomaterials" (ENMs).

It is significant to know that applications of the nanoscale phenomena have already been in vogue for a long time despite a lack of proper understanding of the physical mechanisms involved at the atomic and molecular levels. In this context, the current advances made in developing high-precision measurement techniques combined with theoretical modelling have continuously been striving to unravel more and more about the nature of atomic and molecular species and the complex interactions prevalent in nanoscale materials that are significantly different from those in their bulk counterparts.

Novel properties of the nanoscale materials (i.e., nanomaterials) are particularly found emerging out of two main reasons, namely those associated with a high surface-to-volume ratio that are invariably present in the nanomaterials and that impart higher chemical reactivity, resulting in numerous interactions. The second reason is that the quantum confinement of the involved atomic and molecular species is adjusted to the constrained nanoscale arrangements that are different from bulk and thus dominate in fixing the behaviour of the nanomaterials affecting their optical, electrical, and magnetic properties. Nanomaterials have been produced in one dimension (1d) as nanowires and nanotubes, in two dimensions (2d) as ultrathin films or surface coatings, and in all three dimensions (3d) as nanoparticles (NPs).

1.2.2 Nanotechnology

The definition of nanotechnology includes various aspects like the design, characterization, production, and applications of the structures, devices, and systems by controlling the shape and size of the constituent material building blocks at the nanoscale. Theoretical and experimental investigations carried out in the domain of a large variety of nanoscale materials derived from various sources are subsequently translated into practical applications derived from their engineering and technological considerations for making products that help improve the quality of human life. Considering human beings as an essential component of the complete ecosystem, it becomes mandatory to examine the impact of nanotechnology on the environment as well.

The technological application of nanomaterials from all resources might not necessarily be of use unless the processes involved and the raw materials used are cost-effective and benign to human beings and the environment. Toxicological studies thus form an essential component before finalising the usage of nano scaled materials to solve real-life problems.

1.2.3 Nanomaterials

Nanomaterials must have one or more external dimensions or an internal structure that exhibit novel characteristic properties compared to their counterpart in bulk form. This definition of "nanomaterial" certainly implies retaining the nanoscale features of the material involved in developing macroscale materials, components, and devices. The alternate terminology used in the published literature is "nanostructured materials," wherein the morphological features are expressed more explicitly.

1.3 Physico-chemico-biological Properties of Nanomaterials

The physical features of nanomaterials are generally described in terms of their shapes, sizes, and morphological sub-structures. Nanomaterials are generally available in the form of aerosols, colloidal suspensions, or emulsions. Applications of chemical surfactants are found useful in modifying their surfaces and interfacial properties, including their surface stabilisation against coagulation or aggregation and modifying the outermost layer. It is quite often noted that the nano-size assembly of atomic and molecular species tries to restructure inside as well as at the surface to finally end up in more thermodynamic equilibrium. Surface rearrangements needed for attaining stable morphology might altogether alter the surface atoms to provide more sites for chemical conjugations with chemical species involving strong and weak chemical bonds. The enhanced chemical reactivity of the surfaces of nanomaterials with a large population of unsaturated chemical bonds makes them highly susceptible to interacting with the chemical species that come into contact during synthesis and storage, resulting in a fairly complex situation, including condensation of several species on the surfaces. At the nanoparticle-liquid interface, for instance, polyelectrolytes modify surface properties in response to the interactions between particles and the surrounding environment. They have been used in a wide range of technologies involving adhesion, lubrication, stabilization, and controlled flocculation of the colloidal dispersions.

At some point between the sub-nanometre and the micrometre scale, the simple picture of a nanoparticle as a ball or droplet changes. Both physical and chemical properties are derived from atomic and molecular origins in a complex way. For example, the electronic and optical properties as well as the chemical reactivity of small clusters are completely different from the properties of each component in bulk or at extended surfaces. Complex quantum mechanical models are necessary to predict the evolution of such properties with particle size, and typically very well-defined conditions are needed to compare experiments and theoretical predictions.

In addition to knowing the physico-chemico-biological properties of the individual nanoscale entities involved in constituting a nanocomposite, it is equally important to know the distribution of these nanoscale domains and their interactions via their inter-domain interfaces or intra-phase properties that affect the macroscopic bulk properties through the size or dimension of components and domains via the structural features available at the domain boundaries. There is, therefore, the additional possibility of modifying the surfaces of the embedded domains to influence their overall impact on the material properties available for several applications subsequently. This approach was known centuries ago while developing metal alloys as composites with micro- to nanoscale domain sizes. There is a complex relationship between the structure and the composition of the material, which influences the physico-chemico-biological features of each domain, which may relate to the bulk and surface properties of each ingredient and newly emerging properties localized at the interface.

1.4 Engineered Nanoparticulate Materials

Engineered nanomaterials (ENMs)—intentionally synthesized particulate material species with their characteristic dimensions in the range of 1–100 nm exhibit properties that are different from those of non-nanoscale particles with the same chemical composition. It may be especially emphasized that the properties of NPs are always compared with their corresponding bulk. These ENMs are of special concern due to their adverse impacts on the environment and human healthcare. The basic question that is not yet resolved is whether the characteristic properties of nanomaterials would influence either their exposure or associated hazards in ways that are different from those shown by larger particles of the same composition. The answer to this is still not known. Although nanoparticulate material species are easily taken up by organisms through the processes of ingestion, respiration, or both, with increased residence time and exposure in environmental systems, these effects arise more from their small size than from their unique nanoscale properties. It has been further noted that new fundamental physics or theories beyond those of colloid chemistry are not necessarily needed to describe their interactions with other materials in the biosphere. However, the "non-bulk" properties of NPs in terms of their chemically interactive surface structures might influence processes like dissolution, redox reactions, or the generation of reactive oxygen species (ROS). Such properties, accompanied by corresponding biological effects, would not be produced by larger particles of the same chemical composition. In these cases, new approaches are needed to systematically define nanoparticles and their associated properties in terms of their structural characteristics as a basis for ensuring the reproducibility of the results, identifying underlying mechanisms of toxicity, and predicting environmental behaviour.

Because of their toxicological properties, inorganic metal and metal oxide nanoparticles have sparked significant commercial interest. As a result, it is possible to conclude that a critical size much smaller than 100 nm could be determined experimentally for which these new properties are observed. For mandatory regulation of ENMs, it becomes necessary to define them based on their experimental observations, which indicate that NPs >30 nm do not in general show properties needing regulatory scrutiny beyond those of their bulk counterparts. This critical size, on the other hand, is strongly associated with an exponentially increased number of atoms localised at the surface with decreasing dimensions, delineating a smaller set of NPs with diameters less than 20–30 nm. These smaller NPs have a size-dependent crystallinity that imparts properties quite different from the bulk material. These observations suggest that the toxicological studies of smaller NPs might focus on a smaller set of NPs that show unique nanoscale properties.

NPs <20–30 nm in diameter are characterized by higher surface energy, causing thermodynamic instability. Crystallographic changes involving lattice contraction or deformation, defect generation, rearrangement of the surface atoms, or changes in morphology become necessary for stabilization. Such resultant features are found to affect the interfacial reactivity and the intrinsic properties of the NPs. The size dependence of the optical and electronic properties of quantum dots (QDs) provides a clear example. The fluorescence bandgap starts increasing at diameters <6–8 nm. These properties are attributed to lattice contractions that favour the confinement of electrons and the existence of discrete electronic states that are virtually absent in larger particles. Another example is the size dependence of the thermal properties of nanoparticles. By reducing the diameters of In and Sn-NPs from 100 nm to 10 nm, the melting points can be reduced by 120 °C and 80 °C, respectively, with an exponential drop for sizes 15 nm. Moreover, the normalized heat of fusion behaves in a similar way, whereas it is assumed to be constant according to classical thermodynamics. Electric and magnetic properties are also known to be related to crystallinity and size. For instance, size-dependent changes of the transition temperature like the Curie temperature in ferromagnetic materials.

The introduction of lattice defects and grain boundaries during material synthesis would thus introduce additional features in addition to the changes made in their chemical constitutions. For example, deploying polycrystalline and amorphous phases in nanoscale material samples would also contribute to the observed nanoscale properties accordingly.

The monocrystalline semiconducting materials developed for their specific applications in microelectronics at the nanoscale are thus found to turn into modified crystallinity or even get converted into polycrystalline and amorphous counterparts, offering morphology-specific behaviour that could very well be used as fundamental material building blocks to produce nanocomposite materials. The presence of nanoscale material phases dispersed in a bulk matrix would certainly influence the overall material behaviour in terms of the interfaces formed according to the surface characteristics. This also provides the opportunity to design and produce hierarchical supramolecular structures of various kinds for different applications. The list of such engineered nanomaterials provides an inexhaustive source of nanomaterials that are expected to be developed in the future with the advancement of nanotechnology.

1.5 Chapter Summary

The fundamental concepts of nanoscience have evolved over several decades and are noted to frequently use more than one terminology for understanding the nature of interactions at the atomic and molecular levels, resulting in some confusion for those new to this important area of material science. This is not surprising given that nanoscience has largely been the result of multidisciplinary contributions from various fields. Particularly, the size, shape, and crystallinity of the nanoparticulate material species of inorganic, organic, and biological origin employed in defining the nano-related terminologies might appear to be rather inconvenient to those readers coming from fields other than material science when visualizing the processes before using them in different fields. Keeping this situation in mind, an effort has been made to simplify the terminologies based on the published literature by eliminating the ensuing overlaps. The chance of using different terminologies for a given nanophenomenon is thus minimized while going through the published research papers representing the different fields of application.

Further Reading

National Nanotechnology Initiative (2022). https://www.nano.gov. Accessed 20 Sept 2022

Chapter 2
Nanomaterials and Plant Biomolecules: Basics of Interactions

2.1 Introduction

To find out the fate of nanomaterials and their bioavailability, it is required to know the nature of the interactions that take place at molecular levels that ultimately help in identifying the possible interaction pathways. To choose one specific pathway out of several, it becomes necessary to introduce appropriate ligands via surface functionalization strategies on the surface of the nanocarrier, not only to facilitate the navigation but also to provide them with a unique opportunity to arrive at the desired site with the help of molecular recognition-based chemical conjugation. Once at the targeted sites, the next task is to detach the molecular load using stimuli either externally or available in the external environment. The residues left over after the active ingredients have been off-loaded are subjected to systemic excretion via the system's characteristic routes.

In this generalized strategy of targeted delivery of active ingredients (AIs), it becomes necessary to know the physicochemical properties of the nanocarriers in terms of the chemical reactivity associated with the surfaces of synthesized nanoparticulate species and the surface modifications introduced intentionally to stop their agglomeration or aggregation by coating some stabilizing layer followed by creating newer sites for attaching the loads of AIs, recognition species for attaching chemical moieties for navigation, and imaging, if required, until the nanocarriers involved arrive at the specified locations.

In this whole process, selection of the basic nanoparticulate species is critically important, as the number of choices available might not be viable for commercial exploitation on a large scale. After starting with a simple and rugged nanocarrier system that can be synthesized from a simple, basic form of carbon nanoparticle or other, the feasibility of engineering nanocarriers endowed with desired physicochemical properties emerges. Even more care should be taken to minimize the investment in surface stabilization by following an appropriate synthesis protocol.

A. Nazeer et al., *Targeted Delivery of Nanopesticides and Nanofertilizers in Sustainable Agricultural Farming*, Nanotechnology in the Life Sciences,
https://doi.org/10.1007/978-3-031-41333-9_2

As a general condition, the choice of a basic nanocarrier should be appropriately made of the inorganic, organic, polymeric, and biomolecular families. To minimize the associated toxicity, it is preferred to go for the biopolymeric family of nanocarriers with intrinsic compatibility and minimum toxicity. However, starting with inorganic nanocarriers is not completely ruled out, as their subsequent transformation into ionic species might be used for taking care of unwanted microorganisms as well, and it needs special design of the scheme to minimize the systemic toxicity.

The following discussion is related to the broad area of nanostructured materials and comprises specific topics such as quantum confinement, energy band structure, surface modifications to engineer the physicochemical properties of the nanocarriers, and the interaction of nanomaterials with plant biomolecules leading to characteristic biochemical pathways.

The readers are not overloaded with the long list of publications reporting on each aspect of the discussions included here. Rather, the discussions are built-up using the exhaustive review papers available, and it is assumed that the references used therein are automatically included indirectly through the reviews properly mentioned in the text as well as in the form of further readings. The basic purpose of the chapters included in this book is to make the reader familiar with the current developments and prospects of targeted delivery of nanoscale pesticides and fertilizers in modern crop management to minimize the load of hunger and poverty globally expected by the middle of the current century.

2.2 Nanostructured Materials

It is one of the basic and well-established facts in condensed state physics and material science that the properties of most solids depend upon factors such as chemical compositions, coordination numbers, inter-atomic spacings, and the size of the solid in one, two, and three dimensions. Accordingly, any changes made to any one of these factors are expected to influence the representative properties exhibited by the resultant solid. One of the best-known examples is that of the two extremely different carbon allotropes known as graphite and diamond. The properties of graphite and diamond crystals are quite different. Such property changes have been noted in several other cases as well. So, this statement may be generalized further to say that any solid having a non-equilibrium structural configuration of the constituent atomic species is going to exhibit properties that are naturally different from the bulk solid in thermodynamic equilibrium. It may be noted that the reference for comparison is always an ideal crystalline equilibrium solid.

To take advantage of the modified properties of a solid for its suitable applications, several nonequilibrium structural changes have been made intentionally to study the resultant behaviors. The locations of these nonequilibrium changes could be either near the surface or in bulk. Physical vapor deposition (PVD), chemical vapor deposition (CVD), laser beam interactions, or ion implantation have been extensively used in creating nonequilibrium structures with changed chemical compositions or modified structures for the purpose of surface modifications leading to corrosion resistance, wear resistance, and hardness for their commercial applications. Changes were also incorporated onto a planar solid surface using lithography, direct pattern writing using an STM tip, focused electron and ion beams, and precipitation for producing microelectronic devices and circuits at a large scale, besides realizing quantum dots, quantum wires, or single electron transistors.

While examining the scope of the applications of nonequilibrium solids, it is equally possible to consider them without altering their chemical compositions and densities by quenching them from their molten states. For example, glasses, supersaturated solid solutions, alloys, ion-implanted materials, and materials in a high-compression or tensile state are good examples to be considered in this category.

Another class of materials, known as nanostructured materials, have been extensively explored during the last few decades, representing nonequilibrium states in an entirely different way. The nanostructured materials (NMs) are, therefore, expected to have entirely different properties compared to their bulk counterpart. NMs with a high density of lattice dislocations and point defects, as well as grain boundaries with altered coordination numbers, lattice constants, and chemical composition due to segregation in the case of alloys – all of which result in a nonequilibrium state –It has been noted that the solid samples of nearly 5 nm size have almost 50% of the core as defects. These NMs may therefore be considered as heterogenous compositions composed of the nanoscale crystallites separated by nano size grain boundaries and differing in terms of chemical composition, coordination numbers, and lattice constants from the crystallite cores (Gleiter 1989).

There are, in general, three kinds of microstructures, comprising lamellar, rod-shaped, and equiaxed crystallites. From the composition side, the first kind has all crystallites and interfacial regions of the same chemical composition. The second kind has different chemical compositions in crystallites and their interfaces. The third family has varying chemical compositions across the interfacial regions. In total, there are 12 possibilities in this context of nonequilibrium NMs.

Most of the properties of the solids depend on their microstructures, which consist of their chemical composition, the arrangement of the atoms (the atomic structure), and the size of the solid in one, two, or three dimensions. Accordingly, changes made in these microstructures would result in different properties for the bulk material. One of the well-known examples of the correlation between the atomic structure and the property of a bulk material is the significant variation in the hardness of carbon when it is transformed from graphite to diamond. Comparable variations have also been observed when the atomic structure of a solid deviates from equilibrium or when its size is reduced to a few interatomic spacings in one, two, or three dimensions. Color changes in CdS due to their reduced size in the few nanometers range are good examples in this context (Henglein 1989).

2.3 Quantization Effects

Semiconductor NPs have been found to possess tunable optical and electronic properties because of differences in the quantum confinement effect in terms of the size and shape of the NPs. The size of the NP determines not only its density of states but also restricts the spatial localization of excitons during their electronic excitations. The size adjustment of the NP has been used to manipulate excitonic interactions. For example, the phenomenon of change in the photoluminescence spectra of quantum dots (QDs) is a manifestation of size restriction. The effect of dot size on Auger recombination rates, electron and energy-transfer processes, carrier multiplicity, and dissociation of excitons is still being studied using both experimental and theoretical methods. In addition, NP shape controls the spatial asymmetry of the confinement volume, which leads to shape-dependent effects on exciton behavior. The shape of NP is associated with several different electron-hole (e-h) confinements. Specifically, spherical NPs exhibit 3D quantum confinement, whereas nanoplatelets and nanodiscs exhibit 2D confinement, and nanorods and wires exhibit 1D confinement. For 1D quantum-confined materials, theoretical and experimental studies have shown that the aspect ratios substantially change their optical properties. Nanorods, for example, have been used to enhance carrier multiplicity, optimize resonance energy transfer, and generate linearly polarized light. The properties and shape of 2D quantum-confined materials such as discs and plate like-NPs have also been investigated for studying the excitons confined in the quantum discs to explain the linear optical properties observed. The (e-h) recombination behavior in CdSe nanoplates has been found to differ from that of their corresponding 3D-spherical CdSe-NPs (Scher et al. 2016).

The absorption spectra of CdS-NPs (>6 nm) started showing absorption around 2.4 eV of photon energy (i.e., the band gap of bulk CdS). The absorption threshold, however, started shifting to shorter wavelengths with size reduction, and in the case of size <2.2 nm, the NPs turned colorless. Semiconductors having a band gap smaller than that of CdS exhibit even more drastic color changes, as noted in the case of black cadmium phosphide with a band gap of 0.5 eV, which exhibits all colors of visible light by decreasing the NP size in the range of 2–10 nm. The fluorescence band of the colloids shifted blue as the NP-size decreased. In the case of smaller NPs, the confining 3 eV potential wells with smaller lateral dimensions lead to a quantization of their energy levels. Because the effective mass of an electron in a crystal lattice is significantly smaller than that of a free electron, size quantization occurred more dramatically in low band gap materials with particularly small effective masses of electrons. CdS, ZnO, and PbS are the materials where the shift of the absorption threshold with particle size has been studied in more detail.

Scher et al. (2016) developed a theoretical design called eh-XCHF to understand the detailed changes occurring in the optoelectronic properties of CdSe, CdS, CdTe, and PbS-NP-ellipsoids by performing isovolumetric transformations applicable in a variety of e-h systems, in which the electrons and holes were treated as interacting pairs confined to a three-dimensional parabolic potential. Thus, the eh-XCHF wave function was used in calculating the ground-state exciton binding energy, e-h separation distance, and e-h joint probability of the NPs, and these three properties exhibited strong dependence on ellipsoid shape, even at constant volumes. The exciton-binding energy was greatest in spheres and lowest in rods, which had the highest aspect ratio. The e-h separation distance exhibited the opposite trend. These trends were observed consistently across all semiconductor NPs, regardless of size. Under the independent particle approximation, the overall trends for the investigated optical properties were the same, but the absolute magnitudes were greatly inaccurate.

A crystallite containing fewer atoms disrupts the collective more, and the effective mass increases with NP-size reductions. Size quantization effects have been observed in many systems. Unusual observations were made in thermal and ultrasonic dissolution experiments involving a powder of MoS_2 or WS_2. Heating at 60 °C in acetonitrile for 3 days produced a colloidal solution having 1–3.5 nm NPs with a strongly blue-shifted absorption spectrum in comparison to that of the macrocrystalline material. Similarly, powders of MoS_2 and WSe_2 suspended in water, when exposed to intense ultrasonic treatment, turned out to be transparent, containing quantized NPs of ~3 nm.

2.4 Energy Band Structures

Semiconductor QDs are light-emitting NMs that have been investigated for developing their applications in solar energy conversion, optoelectronic devices, molecular and cellular imaging, and ultrasensitive detections. The enclosure of the electronic charge carriers imparts a unique feature of having discrete energy states in semiconductor nanocrystals called the quantum confinement effect. Consequently, varying the size and shape of these QDs could be used for precise tuning of the energy of discrete electronic energy states assigned to characteristic optical transitions. Further, it facilitates tuning the light emission throughout the ultraviolet, visible, near infrared, and mid-infrared spectral ranges. These QDs, spanning the transition from small molecules to bulk crystals, offer optical properties including carrier multiplication, single-particle blinking, and spectral diffusion. These semiconductors could also be used as building blocks for developing complex nanostructures such as superlattices and multimodal agents for molecular imaging and targeted therapy.

The bulk semiconductors are characterized by a composition-dependent band gap energy (E_g), which is the minimum energy required to excite an electron from the ground state in the valence band into the vacant conduction band. With the absorption of a photon of energy $>E_g$, the excitation of an electron leaves an orbital hole in the valence band. The negatively charged electron and positively charged hole might move in the presence of an electric field to yield a current, but their lowest energy state is an electrostatically bound electron-hole pair, known as the exciton. Relaxation of the excited electron back to the valence band annihilates the exciton, accompanied by the emission of a photon, a process known as radiative recombination.

The exciton has a finite size within the crystal defined by the Bohr exciton diameter (a_B), varying in the range of 1–100 nm depending on the material. In cases where the size of the QD is smaller than the size of the exciton, the charge carriers involved become spatially confined and have higher energy. Therefore, the exciton size delineates the transition between the regime of bulk crystal properties and the quantum confinement regime, in which the optical and electronic properties are dependent on the QD size. Therefore, smaller QDs than a_{B} exhibit size-dependent absorption and fluorescence spectra with discrete electronic transitions.

In the case of CdSe QDs ($E_g = 1.76$ eV, $a_{\mathrm{B}} = 9.6$ nm) can be tuned through quantum confinement to emit fluorescent light throughout the visible spectrum, making them useful for both biological imaging and many types of optoelectronic devices. In addition, the crystalline nature of these QDs imparts a large density of electronic states, yielding giant extinction coefficients and broad absorption spectra that are not available from organic chromophores. State-of-the-art QDs have been engineered to have a quantum efficiency of radiative recombination approaching unity at room temperature, far above what has been achieved from bulk materials. The high efficiency of light emission is largely due to the strong overlap between the electron and hole wave functions in the confined structure, whereas the exciton in bulk semiconductors is not confined in space and can rapidly dissociate, increasing the probability of nonradiative relaxation associated with crystalline defects and charge carriers trapped on crystal surfaces.

Early studies on semiconductor nanocrystals made use of particles embedded in glass matrices, like centuries-old stained glass containing semiconductor crystallite pigments. Today, these nanocrystals are commonly fabricated as colloids suspended in solution or as epitaxial structures grown on solid crystalline substrates. Epitaxial nanocrystals can be reproducibly prepared in a wide range of shapes and sizes in regular patterns and directly incorporated into optoelectronic devices. The solution phase techniques provide exceptional control over size, mono-dispersion, and shape. These particles are typically coated with a monolayer of surfactants and have shown versatile applicability in optoelectronic devices and as biomolecular conjugates. The colloidal semiconductor nanocrystals have largely enabled the physical understanding of size-dependent properties and have catalyzed the broad use of nanometer-sized particles.

The size dependence of the band gap of nanocrystalline semiconductors is the most significant consequence of quantum confinement. By confining the exciton of a semiconductor, the band gap may be tuned to a precise energy depending on the dimensionality and degree of confinement. The shift of the band gap of CdSe nanocrystals is identified as QDs in three dimensions, QW or QR in two dimensions, or QW or discs in one dimension. An increase in the number of confined dimensions yields a stronger degree of electronic confinement and thus a wider range of tunability in the band gap, although exceptions to this trend have been reported for

CdTe nanocrystals. QDs have attracted attention due to their wide optical tunability and utility for biolabeling, whereas elongated structures have been shown to emit linearly polarized light with a wide energy separation between the absorption and emission maxima (Stokes shift), which can reduce light reabsorption for light emission applications. QWs are well-established components of optoelectronic devices, and their colloidal "disk" analogues have recently been reported to have novel piezoelectric applications once confined in polar lattice directions.

The dependence of optical properties on particle size is largely a result of the internal structure of the nanocrystal. However, as the crystal becomes smaller, the number of atoms on its surface increases, which can also impact its optical properties. The atoms on the surface of a crystal facet are incompletely bonded within the crystal lattice, thus disrupting the crystalline periodicity, and leaving one or more "dangling bonds" on each atom pointed outward from the crystal. Most nanocrystals are highly faceted, and each surface contains a periodic array of unpassivated-orbitals with two-dimensional translational symmetry, which may form a band structure like that of the three-dimensional crystal itself. If these surface energy states are within the semiconductor band gap, they can trap charge carriers at the surface, thereby reducing the overlap between the electron and hole and increasing the probability of nonradiative decay events. In practice, however, most semiconductor nanocrystals are not used in vacuum, but are instead either embedded within a solid matrix such as another crystal or a glass or suspended in solution and coated with organic ligands such as trioctylphosphine oxide (TOPO) or hexadecylamine (HDA). Thereby, the dangling bonds on the exposed facets are passivatedby bonding with atoms or molecules, minimizing intra-band gap surface states, and reducing surface atomic reconstruction. For colloidal particle suspensions, molecules of TOPO and HDA adsorb to the nanocrystal surface through dative ligand-metal bonds between the basic moiety on the ligand and metal atoms on the nanocrystal surface, solubilizing the crystal in nonpolar solvents through hydrophobic interactions with the alkyl chains on the ligands. Polar end groups and hydrophilic polymers may be similarly used to solubilize nanocrystals in polar solvents.

The fluorescence emission from QDs can be significantly modulated by alterations to the nanocrystal surface. For example, very small colloidal CdSe nanocrystals often display two fluorescence emission bands, one at the band edge and another at a lower energy resulting from recombination at intra-band gap defect sites on the surface. Deep trap emission arises from the presence of Se^{2-}-rich facets, which poorly bind to most basic ligands, making these nanocrystals especially prone to the surface trapping of holes that primarily reside on the selenium sp^3-orbitals. Indeed, nanocrystals with surfaces terminated mostly by anions typically have little or no fluorescence emission due to the overwhelming number of surface trap states for nonradiative recombination. The addition of excess Cd^{2+} ions can passivate these selenium states, yielding cationic surfaces that can strongly bind to basic ligands and electrostatically shield holes from surface traps. Surface modifications may also introduce defects on the nanocrystal surface. Small amounts of oxidizing agents such as hypochlorous acid or H_2O_2 can etch away fragments of the nanocrystal surface, leaving behind unpassivated orbitals and atomic vacancies and nearly extinguishing the fluorescence with as little as 10 hypochlorous acid molecules per nanocrystal.

The number of surface atoms and the types of facets are also dictated by the nanocrystal shape. The fraction of surface atoms on a CdSe nanocrystal is amenable to different shapes and sizes. Spherical quantum dots have the smallest number of total surface atoms and are thermodynamically the most stable, thus making them ideal candidates for applications in which fluorescence modulation from external stimuli must be minimized. On the other hand, elongated structures such as rods and wires maintain a large fraction of their constituent atoms on their surfaces, making them useful for applications in which the nanocrystal charge carriers can interact with the surrounding environment, such as redox chemistry, energy transfer, photocatalysis, and sensing applications. However, it should be noted that, for nanocrystals of the same volume, the diffusive kinetics decrease much faster for elongated structures, which may detrimentally impact sensing applications in viscous media. Several other interesting material properties are closely linked with the fraction of surface atoms on the nanocrystal. The melting temperature of nanocrystals decreases as the size decreases, as a larger fraction of the atoms in the crystal are on the disordered and incompletely passivated surface. The surface energy differences between different crystal phases have also been used to explain alterations to the pressure-induced phase transitions in high-surface-area semiconductor nanocrystals relative to their bulk counterparts.

For light-emitting applications, it is helpful to coat the QDs with an insulating inorganic shell to stabilize and maximize fluorescence. This not only passivates the surface bonds but also buries the semiconductor in a potential energy well, concentrating the charge carriers in the core and away from the surface. Thereby, surface defect states and trap sites exert no impact on the fluorescence efficiency. In solid-state devices, this process is quite simple, as the overgrowth of an inorganic layer with a wider band gap is realized with complete surface coverage. In colloids, the task is more challenging, but major progress is being witnessed in colloidal QDs like (CdSe) ZnS and (CdSe) CdS with efficient and stable fluorescence. Wider band gap CdS and ZnS shells not only isolate the cores, but S^{2-} has a much lower oxidation potential than Se^{2-}, resulting in a higher threshold to photo-oxidative degradation and surface defect formation. Additionally, the implementation of shells composed of elements with smaller atomic radii is crucial for ligand binding, as a higher packing density of organic ligands can enhance colloidal stability and create a steric barrier against smaller molecules.

Unlike solid-state crystals, small-molecule chromophores, and fluorophores, QDs exhibit unique optical properties controlled by their structure and size. One of the unique features is the random on-and-off blinking phenomenon exhibited by the single QD. Illumination at higher energies might result in the production of several charge carriers in a single particle, which may increase the efficiency of photovoltaic and light-emitting devices.

The blinking phenomenon in QD is associated with either repeated ionization of the nanocrystal due to the Auger process or the trapping of one of the charge carriers at or near the particle surface. Consequently, the off-time duration can be reduced in low-dielectric media that poorly solvate the expelled charges, and through the adsorption of thiolate ligands and amines, which can neutralize surface traps for charge carriers. Growing an insulating shell around the nanocrystal results in a deeper and wider potential well with shorter off-time durations. Recently, it was shown that colloidal (CdSe) CdS-QDs with thick insulating shells turned out to be almost nonblinking and resembled the self-assembled QDs embedded in crystalline substrates. Blinking could be eliminated in core-shell structures, in which there is a smooth composition gradient from the core to the shell. This is so because more gradual changes in the confinement potential prevent the Auger recombination that allows the fluorescence emission from the ionized nanocrystals. However, the exact structure of these nonblinking QDs and their light emission mechanisms are not yet clearly known.

Carrier-multiplication is another attribute of QDs. In a QD excited by a photon with energy more than twice its band gap, the electron and hole generated initially might release the excess kinetic energy by creating a secondary electron-hole pair in the form of a biexciton, resulting in internal quantum efficiency exceeding 100%. Since the excess kinetic energy of the primary pair is not dissipated in the lattice vibrations, the improved efficiencies become useful in PV devices. These processes are efficient in QDs compared to their bulk counterparts, due to the suppression of the phonon-assisted decay rate and the large energy separations between electronic energy levels. However, slower auger transitions might eliminate the excess excitons, such that they do not participate in radiative recombination. Recently, biexciton and triexciton fluorescence phenomena were observed at room temperature with higher quantum efficiency and long lifetimes in highly luminescent (CdSe) ZnS-QDs. The current challenge is to harness these extra charge carriers by maximizing their stabilization before they decay via Auger processes to convert them into an external current, possibly using a weakly coupled multilayer structure wherein the excitons can segregate, as charge transfer in semiconductor nanocrystals generally occurs much faster (~500 fs) than Auger processes.

Quantum confinement shifts the band gap of most semiconductors more than 1 eV, giving an enormous range of continuous tunability through size and shape in a single configuration, in addition to enabling independent tuning of size and band gap through the implementation of alloyed materials like $CdSe_yTe_{1-y}$, and $Cd_xZn_{1-x}S$. Lower energy light emissions from defect states to the ground state are enabled by chemical impurity doping that creates an intra-band electronic energy level. Doped nanocrystals might possess properties useable in biolabeling and device applications based on large Stokes shifts, paramagnetic properties, and improved lasing. Presently, however, the synthesis of these doped nanocrystals is still a challenge due to the chemical dissimilarities between the dopants and their crystalline matrices. More recently, the manipulation of nanocrystal heterostructures has been found to provide complex control over charge carrier wave functions, resulting in newer optical properties.

Core-shell semiconductor heterostructures were reported in 2003, in which the conduction and valence bands of the core and shell materials were staggered, resulting in the segregation of the electron and the hole between the core and shell materials. For example, core-shell (CdTe) CdSe particles have a minimum conduction band energy in the CdSe shell, whereas the maximum energy of the valence band is in the CdTe core. These energy band offsets segregate the electron from the shell and the hole from the core, and carrier recombination can occur across the interface at a lower energy than the band gaps of either of the constituent materials. During shell growth, the wavelength of emission shifts strongly to the red, adding a new dimension to band gap engineering. The reduced spatial overlap between the electron and hole also results in a major reduction in band edge oscillator strength and a significant increase in excited state lifetimes. This band alignment is designated as "type-II" to distinguish it from "type-I" band alignments like those of (CdSe) ZnS in which the electron and hole overlap in space in the core material. Type-II materials are expected to be beneficial for photovoltaic devices, especially with the development of anisotropic materials in which the segregation of charges is directional, which may enhance directional charge transport in future electronic devices. The use of type-II QDs offers independent control over which charge carrier is accessible to the surface for charge transfer applications. It is also possible to sequester both charge carriers in the shell material in inverted "type I" alignments, which resemble a quantum well in which one-dimensional confinement is present in the radial direction. The capacity to transition between core and shell localization has recently been exploited to tune the interactions between charge carriers and core-localized magnetic atoms of (ZnSe:Mn)CdSe QDs.

The presence of strain in semiconductor QDs provides another latitude for modifying their characteristic properties. Lattice strain has been found responsible for growing QD islands on lattice mismatched solid substrates in the Stranski-Krastanov growth mode, introducing substantial strain in the nanocrystals. Colloidal core-shell QDs (CdSe) ZnS with a large mismatch in bond length between the core and shell materials (12%) result in a strained nanocrystal. Growth of a material on a substrate with a different bond length can only proceed to a critical thickness before a lattice defect (commonly a misfit dislocation) will appear near the interface to ease out the strain, allowing the materials to relax to their native bond lengths and introducing carrier traps in the process. However, colloidal QDs behave differently from those fabricated on solid substrates because, unlike large crystals, an isolated QD can deform to adapt to the overgrowth of a straining shell, and the large surface area allows the stress to be distributed over a large fraction of the constituent atoms. Deformed semiconductors force the material to adopt a nonequilibrium bond length followed by an altered band gap. This effect can be observed directly using a diamond anvil cell on a bulk crystal, demonstrating a linear relationship between the band gap and pressure, which also holds true for semiconductor QDs.

A recent study emphasized the important role of strain in colloidal (core) shell QD-heterostructures. For QD-cores smaller than a specific size, the critical thickness for a particular shell could be of any value without the introduction of lattice-relaxing defects. For example, the bond length difference between CdTe and CdS is 11.4%, but CdTe cores <4 nm the growth of CdS shell is purely through coherent growth without defect formation. Thus, if the materials composing the heterostructure are reasonably deformable, strain can be tolerated and shared across the entire QD without the introduction of defects. Because such heterostructures can be grown coherently as a single crystalline domain, homogeneous lattice strain results in significant bond length alterations that can modify electronic band structure. Such a compressive strain will shift both the conduction and valence bands to higher energy, and a tensile strain will shift the bands to lower energy. Accordingly, the coherent growth of compressive shells like ZnSe or CdS on a CdTe core in colloidal form would change the relative energy bands in the heterostructure, causing "typeII" band alignments with widely tunable band gaps. This effect is also expected to enhance the 'type-II' nature of strained heterostructures such as (CdTe) CdSe that are already "typeII" in bulk. For this type of band gap engineering, an important component is the implementation of CdTe QD cores. In comparison to commonly investigated CdSe, CdTe is softer, has a larger deformation potential, and has a much larger difference in lattice constant compared to common shell materials such as ZnS and CdS. Furthermore, similar attributes are expected in the III-V antimonides and the other II-VI tellurides.

Strain also plays a critical role in bending charge carrier wave functions at heterostructural interfaces. Because the strain field in a lattice-mismatched (core) shell QD is inhomogeneous in the shell, the band edge is thought to deform radially, resulting in inhomogeneous charge carrier distributions. This is likely the cause of the different fluorescence behaviors of (CdSe) CdTe and (CdSe) ZnTe QDs. Both heterostructures are "type II," with the hole sequestered in the shell and the electron in the core, but heterostructures with CdTe shells have quantum yields (20–40%) that are dramatically larger than those (<1%) with ZnTe shells. It is important to realize that the lattice mismatch between CdSe and CdTe is 7%, whereas the lattice mismatch is only 0.6% for ZnTe. The high strain in the CdTe shell will shift the hole wave function toward the core material, increasing the overlap between the electron and hole and thus the probability of carrier recombination. The near absence of strain in ZnTe shells substantially increases the wave function localization at the QD surface, where surface traps increase the probability of nonradiative relaxation. Indeed, in true unstrained "type II" band alignments, the probability of band-edge recombination efficiency should be very low, and engineering the strain in such heterostructures may be important for modulating recombination versus carrier separation. Inhomogeneous strain fields in shells have also been proposed as how Mn^{2+} dopant energy levels can be continuously tuned in the shells of (CdS) ZnS nanocrystals.

Major advances are anticipated in fundamental studies and practical applications such as semiconductor QDs. For instance, the synthesis of new QDs with unusual structures and properties is almost inexhaustible and will continue to exhibit doped and strain tuned QDs. Through single-particle fluorescence, extinction, and tunnelling spectroscopies as well as ultrafast spectroscopies, it is possible that complex electronic states arising out of defect and surface states could soon be understood for various types of semiconductor QDs. There are also a wide variety of new nanocrystalline materials available with a diverse range of chemical, elastic, and optical properties. Oxide materials like ZnO would be an exceptional shell material for QD capping due to a wide band gap and resistance to oxidative degradation; IV-VI semiconductors have uniquely positive deformation potentials; and mercury-based II-VI materials may allow the continuous tuning of band gaps through spontaneous cation exchange reactions. For PVSC applications, it will be important to engineer QDs for multiexciton generation and efficient charge carrier separation. For biomedical applications, it is important to minimize the overall size of bio-conjugated nanocrystals that would have reduced steric hindrances and nonspecific protein adsorptions, be useful for synthesizing chemically activatable or photo-switchable QDs for super-resolution optical microscopy and understand the potential toxic effects of semiconducting NMs.

2.5 Surface Modifications

NPs are known to have highly chemically active surfaces with a higher density of unsaturated dangling bonds that impart an inherent tendency to adsorb the charged species present in the surrounding environment and act as traps for electrons and holes. The photo-generated charge carrier absorption is strongly dependent on the existence of these traps. Fluorescence experiments demonstrated this by exhibiting different fluorescence spectra of CdS-NPs before and after surface modification. CdS-NPs in colloidal solution, prepared by adding H_2S to a cadmium salt solution, were found to have a very weak red fluorescence, peaking at a photon energy about 0.4 eV below the absorption threshold. This fluorescence was attributed to the radiative recombination of trapped charge carriers competing with the radiation-less recombination as the dominant process. After surface modification by adding 300% Cd^{2+} ions and increasing pH to 11, a bright green-blue fluorescence was observed with a quantum yield exceeding 50%. The maximum of this fluorescence was found at the absorption threshold while dealing with band gap recombination of the free charge carriers. The effect was assigned to the formation of the cadmium hydroxide layer, where Cd^{2+} ions were coordinatively bound to surface sulphide anions to form ($S^{2-}Cd^{2+}0H^{-}$) structures, which did not allow the surface imperfections to participate in the trapping of the charge carriers. The CdS-NPs with a mean diameter ~6 nm before surface modification became substantially larger due to the presence of a cadmium hydroxide layer, as confirmed in dynamic light scattering experiments. The stability of these NPs against both light and OH radical attack was also found to increase by several orders of magnitude.

Further studies conducted using charged ligand molecules with different affinities showed that the ligands bound to the NP surface not only controlled the growth of the NPs during synthesis but also prevented their aggregation. Inter-particle repulsive force, steric hinderance, or a hydration layer on the surface combined together could produce stable NPs with an appropriate core and dispersion medium. First of all, the ligands should be bound to the NP surface by chemisorption, electrostatic attraction, or hydrophobic interaction, usually offered by the head group of the ligand molecule. Several functional groups have affinity for inorganic surfaces such as thiol and gold. The interaction of the polar or charged ligand molecules provides solubility in aqueous solvents, while NPs with hydrocarbon chains are only soluble in organic solvents like hexane, toluene, and chloroform. Certain amphiphilic ligands like poly(ethylene) glycol (PEG) when attached to NPs along with similar or other ligand molecules make the NPs soluble in a number of solvents. In organic solvents, the NPs covered by hydrophobic ligands are prevented from aggregation. Bonds between the inorganic NP-surface and, an electron-donating end-group of ligand molecules like thiol, amine and phosphine, establish a competitive dynamic binding/unbinding that may cause the ligands to get off by excessive washing or mass action by another incoming ligand, which might compromise the stability of the NPs resulting into aggregation and precipitation. In fluorescent QDs, light irradiation can enhance oxidation of the inorganic NP surface, and photo-oxidation eventually resulting in aggregation caused by desorption of the stabilizing ligands.

In aqueous solutions, with the ligand-NP interaction being basically the same, a number of different effects might be responsible for affecting stability. Normally, hydrophilic NPs are stabilized by electrostatic repulsion experienced by equally charged ligands on the NP surface. In the presence of high salt concentrations, however, due to screening, the NPs can come close to each other until the attractive forces comprised of induced dipole interaction, van der Waals bonding, or hydrogen bonds eventually favour the NPs to agglomerate. Depending on the isoelectric point and the pH of the solution, NPs can also lose or change the sign of their charge. NPs are often complex entities with properties different from simple model systems. NPs are, for instance, not spherical hard objects but are covered with a soft organic ligand shell, of which the charge distribution is not precisely known. Depending on the external factors, heterogeneous surface coverage may result in hydrophobic patches on the NP-surface, or the ligand shell may undergo conformational changes. Furthermore, bi- or multi-valent, oppositely charged ions or polyelectrolytes could be used to electrostatically bridge the NPs, resulting in aggregation.

Strongly binding molecules forming a dense layer stabilize NPs better than weakly binding ones, in particular in further processing and purification steps after the synthesis. In aqueous solutions, strongly charged ligand molecules like carboxylic or sulphonic acid groups are found to stabilize the NPs for longer times and at higher salt concentrations. The ligand molecules providing steric stabilization are found to be much more resistant to high salt concentrations than electrostatically stabilized NPs, provided they are strongly bound to the NP surface, while poor stability is noted in cases of NPs that are not stabilized by ligand molecules.

To improve the stability of a given NP, the ligand molecules can be exchanged with others that could impart new properties or functionality. In most cases, the incoming ligand molecule binds more strongly to the inorganic NP surface. Au-NPs in an aqueous solution synthesized by citrate reduction have negatively charged citrate ions adsorbed on their surfaces, contributing to their stabilization by electrostatic repulsions. Such colloids may be stable for years as synthesized solutions. They cannot, however, be concentrated, as they aggregate irreversibly. The citrate layer can be replaced by ligands binding stronger to the NP-surface such as sulphonated phosphines or mercaptocarboxylic acids, and mercaptoacetic acid (MAA), mercaptopropionic acid (MPA) or mercaptoundecanoic acid (MUA). Modifying the NPs with phosphines allows for the creation of highly concentrated NP-solutions; the NPs can be precipitated by salt-induced aggregation and redissolved again as single NPs in low-salt buffers. Then, thiol-containing ligand molecules can be added to replace the phosphine, a strategy commonly employed for the attachment of thiol-modified DNA to NPs. In the case of incompletely saturated NPs, the remaining phosphine molecules covering the surface help stabilize the NPs. In organic solutions, Au-NPs are commonly synthesized by the Brust two-phase method, which employs tetraoctylammonium bromide and dodecanethiol as stabilizing ligands that also determine the NP's size. Alternatively, the NPs can also be synthesized in the absence of the thiol ligand and solely with the organic quaternary ammonium ion. In this way, synthesis and final capping of the NPs are two independent steps in the synthesis, and the commonly used dodecanethiol can be substituted by a variety of other molecules in order to vary the surface properties and functionalization of the NPs. Similarly, Au and Ag-NPs of different sizes are synthesized using oleylamine in toluene followed by thiol derivatization. Thiol groups are considered to possess the highest affinity to noble metal surfaces, and in particular to gold. While this binding is often termed "chemisorption", sometimes also noted as a "covalent bond," the exact processes and the microscopic nature are still subjects of research and discussion.

Contrary to the case of self-assembled monolayers (SAMs) on a well-defined planar crystal face, the surface of a NP consists of not only a number of different crystal facets but also a large number of edges, terraces, and vertices, resulting in binding sites with different affinities for the ligand molecules. This complicates the characterization of the ligand shell compared with "classical" SAMs, like the well-studied Au-alkanethiol system. In addition, bound ligands appear to be mobile on the surface due to diffusion on the NP-surface after getting attached. In the aqueous phase, the replacement of ligand molecules may be facilitated by additional detergents, and the incoming, stronger binding ligands may render the NPs more stable with regard to possible aggregation at high salt concentrations or an acidic pH. Ligands with two thiol moieties have also been reported. The same principle holds for QDs, CdSe, and CdSe/ZnS, among others, that are often stabilized by trioctylphosphine (TOP) or its oxide (trioctylphosphine, TOPO), which bind preferentially to the Cd or Zn atoms. Also in these systems, binding dynamics and surface diffusion are found; the surface coverage may depend on the particle size and geometric shape of the ligand molecules involved. Besides, hexadecylamine or other amines are also commonly used for CdSe QDs, which can also be replaced by stronger binding molecules like one or more thiol groups. In the case of QDs, the ligand shell may influence the fluorescence properties of the NPs, in particular the quantum yield, especially in the aqueous phase. Furthermore, ligand exchange was also demonstrated with dendrimers and thiol-containing peptides in the case of Au-NPs, as well as with proteins binding to QDs by replacing MAA on the surface. The same principles also apply to Au-rods, where the replacement of the cetyltrimethylammonium bromide (CTAB) double layer by thiol-containing PEG was reported. These considerations imply that, for ligand exchange, the new ligand molecules should have an affinity as strong as possible for the inorganic core in order to quickly and effectively replace the original surfactant molecules. In addition, the molecular geometry of the ligands in relation to the particle diameter is a factor that influences how densely the molecules are packed around the particles, which in turn influences ultimately the colloidal stability of the NPs. However, ligand-coated NPs differ from simple micelles consisting of the ligand molecules alone, which are held together only by intramolecular forces on NPs. The ligand molecules are additionally attached to the NPs' surface mostly via a chemical functional group, which can be electrostatic for gold NPs capped with tetraoctylammonium bromide or covalent like a gold-thiol bond. Naturally, ligand molecules that are strongly bound to the NP-surface or more tightly to each other will be less likely to get off the NPs' surface, as shown in different peptide sequences. In analogy to the concept of critical micelle concentration, such ligand molecules bind dynamically to the NP surface and are thus subject to mass action and may be washed off by continued purification of the NPs.

Polymer coatings provide an alternative to small-molecule ligands, imparting the properties of macromolecular systems to the NP-surface. PEGylated-NPs facilitate passive targeting of tumour tissue through the EPR effect. This stealth coating with PEG and other polymers prevents the adsorption of blood serum proteins, increases circulation time, and enhances the probability of NP-permeation into tumour tissues. Attaching targeting moieties to PEGylated NPs further enhances their tumour targeting. Recently, PEGylated A-NPs were decorated with various amounts of human transferrin (Tf), generating a series of Tf-coated NPs with near-constant size and electrokinetic potentials, and delivered into mice bearing Neuro2A tumours. PEGylation demonstrated enhanced stabilization in salt solution compared to unmodified particles. A Tf-content dependent NP-accumulation in solid tumour was clearly visualized *in vivo*. Polymer coating is a versatile process, allowing the core properties of different NMs to be applied. In one example, a cooperative nanosystem consisting of two discrete NMs was shown to be a potential therapeutic system. The first component was a PEG-coated gold nanorod (Au-NR) that accumulated in murine tumour tissues by the passive targeting mechanism. These NRs were used as photothermal antennas to heat the tumour tissues via remote infrared irradiation. Local tumour heating accelerated the recruitment of the second component: a targeted NP consisting of either magnetic nanoworms or doxorubicin-loaded liposomes. The targeting species was a cyclic peptide that binds to the stress-related protein p32 that is upregulated in tumour cells upon thermal treatment. Treatment of mice containing xenografted tumours with the combined NP-cooperative systems resulted in a significant reduction of tumour volume compared with treatment with individual nanoparticles. An extended study was carried out using similar PEG-Au-NRs but with different targeting NPs that exhibited improved *in vivo* tumour targeting efficiency. Au-NRs were pre-delivered to the tumour site through passive targeting and subsequently irradiated with a NIR laser to damage blood vessels that activated coagulating proteins. By functionalizing the surfaces of the NWs and liposomes with coagulating cascade targeting peptides, an increase in particle accumulation in tumours was observed. Polymer-functionalized NPs facilitate the controlled release of drugs as well. For example, gold nanocages were functionalized with poly(N-isopropylacrylamide) (pNIPAAm) and derivatives that change conformation in response to small variations in temperature. When exposed to NIR, the temperature increase caused the polymer chains to collapse, exposing the pores of the nanocages and thereby releasing the anti-cancer drugs. When the heating was turned off, the drop in temperature brought the polymer back to its original conformation, closing the pores and hence stopping the release. These nanocages were used to control the dosage release by regulating the power density and irradiation time, as discussed earlier (Mout et al. 2012).

Biomolecule-coated NPs provide specific attributes that are difficult to achieve using synthetic materials, such as efficient delivery of biomacromolecules with minimal cytotoxicity. In one of the early examples of bio-nanotechnology, the oligonucleotide-Au-NP conjugates were produced through gold-thiol bond formation. Subsequent studies demonstrated that DNA-Au-NP complexes were capable of inducing gene silencing via an antisense mechanism. A notable feature of DNA-Au-NPs compared to free DNA molecules is their efficient internalization into cells. This rather surprising uptake of anionic NMs is apparently mediated by scavenger receptors. Several research groups have also reported functionalized Au-NPs as siRNA delivery agents. Gene regulation with polyvalent siRNA-Au-NP conjugates was reported, where siRNA was covalently coupled with Au-NP through an ethylene glycol spacer and an alkylthiol, resulting in a serum-stable conjugate that made it useful material for *in vivo* gene regulation. Another advantage of attaching siRNA to Au-NPs is that the resulting NPs provoke a less robust innate immune response compared to Lipofectamine, a commercial transfection agent. The functionalization of NPs with oligonucleotides opened up a variety of new opportunities in molecular diagnostics, including the detection of nucleic acids and intracellular biomolecules like ATP. In another example, novel oligonucleotide-Au-NP probes were developed by a hybridizing fluorophore-labelled complementary sequence and used as cellular "nanoflare" for visualizing and quantifying mRNA in living cells. In this scheme, Au-NPs were functionalized with thiolated oligonucleotides complementary to a specific mRNA transcript and hybridized with a short cyanine (Cy5) dye-terminated reporter sequence. When the reporter flare sequence was bound to the NP, the Cy5 fluorescence was quenched due to its proximity to the Au-NP surface. Once the nanoflare is delivered into cells, the target RNA displaces the reporter flare strand, generating a "turn on" fluorescence signal. These probes do not require any microinjection or additional transfection reagents to enter the cells and are highly resistant to enzymatic degradation as well as nontoxic under physiological conditions. In another approach, Au-NPs functionalized with fluorophore-labelled aptamers were used to determine the level of intracellular ATP by a similar mechanism. NPs coated with proteins and antibodies can bind to cell surface receptors, providing targeted delivery. For example, multivalent gold and silver NPs were engineered by loading Herceptin, an antibody to the Her-2 receptor overexpressed in ovarian and breast tumour cells. The avidity between the NPs coated with Herceptin and its receptor Her-2 targeted them into the tumours.

2.6 NMs and Plant Biomolecules

Although the recent progress of nanotechnology in biomedical sciences has already had a significant impact in the fields of therapeutic and diagnostic applications, the comprehension of NP-plant interactions, including uptake, mobilization, and accumulation, is still in its early stages. Deeper studies are needed to establish the impact of NMs on plant growth and agro-ecosystems while developing nanotechnology applications that enable smart crop management.

The understanding of various mechanisms like uptake, translocation, and accumulation of NMs, along with the assessment of their adverse effects on plant growth and development, requires an accurate evaluation of NP-plant interactions. Plant uptake of NPs is difficult to predict as several factors related to the NP are involved in terms of size, chemical composition, net charge, and surface functionalization. Similarly, their application routes and interactions with the environment via soil texture, water availability, and microbiota; the constraints due to the presence of a cell wall; and the physiology and multifaceted anatomy of individual plant species. Most of the previous studies have dealt with the uptake of small metal and metal oxide NPs, due to their wide uses in industry and the ease of their detection and tracking by microscopy techniques. However, compared to the great wealth of information available in metazoans, only a handful of integrated comparative analyses were able to establish the contribution of physicochemical features like size, charge, coatings, and other features of NPs in plant-NP interactions.

Engineered NMs can be better applied either to the roots or to the vegetative part of plants, preferentially to the leaves. At the shooting surface, NPs can be passively taken up through natural plant openings with nano- or microscale exclusion size, such as stomata, hydathodes, stigma, and bark texture. However, additional plant anatomical and physiological aspects need to be considered for a better understanding of the dynamics of NP-plant interactions. For instance, shoot surfaces are generally covered by a cuticle made of cutin and cutan biopolymers and the associated waxes, which function as a lipophilic barrier to protect above-ground plant primary organs, leaving access only through natural openings. The dynamics of NPs at the cuticle level are poorly investigated, but at present, this barrier appears to be almost impenetrable to the NPs, although TiO_2-NPs are found to produce holes in the cuticle. Trichomes on plant organs can affect dynamics at the plant surface by entrapping NPs on the plant surface and thus increasing the duration of the permanence of the exogenous materials on tissues. Damages and wounds inflicted may also function as viable routes for NP internalization in plants in both aerial and hypogeal parts. Delivery methods also seem to influence NP-uptake efficiency in the plants.

The aerosol application promoting the higher internalization rates of different NPs was reported recently with respect to NP drop cast in watermelon. Also, leaf lamina infiltration might be used to force NM penetration in plant tissues, as reported for SWCNTs affecting gene delivery. At the root level, rhizodermis lateral root junctions may provide easy access to NMs, especially near the root tip, while the upper parts are impermeable due to the presence of suberin. Generally, the dynamics of NP uptake appear to be more complex in the soil compared to the plant aerial part. Several factors like the presence of mucilage and exudates, symbiotic organisms, and soil organic matter may influence NPs' availability. For instance, root mucilage and exudates normally excreted into the rhizosphere play a dual role of promoting NP adhesion to the root surface, which in turn may enhance NP-internalization rate or, conversely, these gel-like substances may also trigger NP-trapping and aggregation. Recent observations using X-ray computed nanotomography and enhanced dark-field microscopy combined with hyperspectral imaging, have demonstrated that root border cells and associated mucilage tend to trap gold nanoparticles irrespective of particle charge, while negatively charged NPs are not sequestered by the mucilage of the *Arabidopsis thaliana* root cap and translocate directly into the root tissue.

Symbiotic bacteria and fungi in the soil are found playing controversial roles as well; though in general, they enhance the accumulation of different types of heavy metal nanoparticles in true grasses but reduce nano-Ag and nano-FeO uptake in legumes (Sanzari et al. 2019).

2.7 NP Transport

Once penetrated the plant's outer protective layers, regardless of aerial or hypogeal exposure, NMs have two routes in the plant known as apoplastic and symplastic paths. Apoplastic transport occurs outside the plasma membrane through the cell wall and extracellular spaces, whereas symplastic movements involve the transport of water and solutes between the cytoplasm of adjacent cells connected by plasmodesmata and sieve plate pores.

Apoplastic transport has been demonstrated to promote radial movement of NMs, which may move NPs to the root central cylinder and the vascular tissues and promote their movement upwards in the aerial part. This NP-translocation method is useful for applications that require systemic NP-delivery. However, the Casparian strip, a longitudinally oriented layer made of lignin-like structures, prevents the completion of this radial movement in the root endodermis. To bypass this natural barrier, water, and another solute switch from the apoplastic to the simplastic path. Similar abilities to circumvent the block at Casparianstrip have been documented for different kinds of NPs, as reviewed earlier. This may happen especially in those anatomical regions where the Casparian strip is not yet properly formed, such as root tips and root lateral junctions.

The symplastic transport of NPs requires that, at some point, they penetrate inside the cells. The presence of a rigid plant cell wall creates a physical barrier to the cell's entry and makes the intracellular delivery of NPs in plants much more difficult than in animal cells. Basically, the cell wall is a multi-layered framework of primarily cellulose/hemicellulose microfibrils and scaffold proteins, creating a porous milieu that acts as a narrow selective filter with a mean diameter <10 nm, with some exceptions up to 20 nm. This is a critical point and currently represents one of the main hurdles to the design and implementation of bioengineering tools in plants. However, different types of NPs with a mean diameter between 3 and 50 nm and CNTs have been demonstrated to easily pass through the cell wall in many plant species.

Subsequent cell internalization may occur preferentially by endocytosis, although alternative cell entry mechanisms, such as those based on pore formation, membrane translocation, or carrier proteins, which have already been described in cells and in invertebrate models, need to be further elucidated in plant cells. It has been demonstrated that MWCNTs may enter *Catharanthus roseus* protoplasts via an endosome-escaping uptake mode.

Once in the cytoplasm, inter-cellular movements of NPs are facilitated by plasmodesmata, membrane-lined cytoplasmic bridges with a flexible diameter (20–50 nm), which ensure membrane and cytoplasmic continuity among cells throughout plant tissues. Transport of NPs with variable sizes through plasmodesmata has been described in *Arabidopsis*, rice, and several other plant species.

Through the symplastic and apoplastic pathways, small NPs can reach the xylem and phloem vessels and translocate in the whole plant to different tissues and organs like flowers, fruits, and seeds with a strong capability to import fluids from the phloem and tend to accumulate NMs. Besides plant toxicity, NP accumulation in specialized organs raises another important issue related to their safe use in human and animal consumption.

Studies in different crops, such as maize, spinach, and cabbage, reported the ability of metal-NPs to penetrate seeds and translocate into the seedlings, without significant effects on seed viability, germination rate, or shoot development. These data suggest the possible use of functional NPs for seed priming, plant growth stimulation, and in limiting environmental conditions.

2.8 NP Phytotoxicity

The understanding of NM toxicity in crop plants is currently at a very early stage, but it is important for the implementation of innovative agro-nanotech tools and products. Current studies in plants have studied particularly unrealistic cases comprising short-term and high-dose exposure in model media and plant species and observed contradictory results. Most of the studies have shown that in cultivated species like tomato, wheat, onion, and zucchini, excess metal-based NPs cause an oxidative burst by interfering with the electron transport chain and impairing the ROS detoxifying machinery, adversely affecting genotoxicity, secondary metabolism, hormonal balance, and growth. Recent transcriptome analyses revealed that exposure to different types of nanoparticles (NPs) of zinc oxide, fullerene soot, or titanium dioxide repressed a significant number of genes involved in phosphate-starvation, pathogen, and stress responses, with possible adverse effects on plant root development and defense mechanisms in *A. thaliana*. A recent approach, including omics data from tobacco, rice, rocket salad, wheat, and kidney beans, confirmed that metal-NMs lead a generalized stress response, with the prevalence of oxidative stress. These data suggested that further studies based on high-throughput analysis of genetic and metabolic responses triggered by NP exposure would be needed to shed adequate light on many aspects of NP phytotoxicity in crops, even in the absence of overt toxicity at the phenotypic level. Considering this evidence, it appears possible to exploit for future applications in plants engineered-NMs for which a safe profile has been already established in animal systems, such as soft polymeric NPs (Sanzari et al. 2019).

2.9 NPs and Plant Physiology

The main plant physiological indices of the toxic effects of NPs are the germination percentage, root elongation, biomass, and leaf number. NPs can have substantial negative effects, such as a reduction in seed germination and suppression of plant elongation and can even cause plant death. Several previous plant nanotoxicity studies have determined the inhibition of plant species such as soybean, maize, wheat (*Triticum aestivum*), ryegrass, and barley by exposure to NPs (MWCNTs, single-wall carbon nanotubes, ZnO-NPs, Ag-NPs, and Fe-NPs); several aspects of plant growth were affected, including seed germination, shoot length, biomass, and gene expression. Growth inhibition was observed when *Bacillus thuringiensis* (Bt)-transgenic cotton was exposed to SiO_2-NPs. When wheat plants were grown in a sand matrix, CuO-NPs inhibited their growth and changed the structure of the roots. The CuO-NPs significantly reduced the fresh weights and root lengths of *Arabidopsis* seedlings and the germination rate and biomass of rice seeds.

The rare earth oxide NPs (CeO_2, La_2O_3, Gd_2O_3 and Yb_2O_3) had a detrimental effect on plant growth in radish, tomato, rape, lettuce, wheat, cabbage, cucumber, and corn when added to roots at high concentrations. TiO_2-NPs increased the content of total chlorophyll and catalase (CAT) and decreased ascorbate peroxidase (APX) content in leaves (Yang et al. 2017).

2.9.1 NPs and Plant Hormones

Plant hormones are active organic materials that are produced by plant metabolism and can regulate physiological responses during plant growth and mediate responses to challenges. The content and activity of plant hormones are considered important indices of toxicity in plants. The CeO_2-NPs had no significant effect on indole-3-acetic acid (IAA), abscisic acid (ABA), and gibberellic acid (GA) in the leaves of Bt-transgenic and conventional cotton compared with the control group. When conventional cotton was exposed to 500 mg/L CeO_2-NPs, the content of trans-zeatin-riboside (t-ZR) in the leaves decreased by 25% when compared with the control group alone. The IAA and ABA content in the roots of transgenic and non-transgenic rice was increased in response to γFe_2O_3. NPs had a significant influence on the production of plant hormones. The decreases in phytohormone concentrations were evident in rice seedlings upon exposure to carbon nanotubes. If Ag ions inhibited the production of ethylene, the interaction between IAA and ethylene would be substantially weakened (Yang et al. 2017).

2.10 NPs and Crop Quality

Previous studies of hydroponic plants have shown that the accumulation of NPs in the environment could profoundly change the quality and yield of soil-based food crops. Protein was found to be insensitive to stimulation by Ag-NPs when compared with carbohydrates, with protein content increasing only at high Ag-NP concentrations (100 mg/L). The ZnO-NPs increased the starch and protein content and decreased the micronutrient (Cu and Mo) concentrations in cucumber. Rice treated with $n$$CeO_2$ had less Fe, S, prolamin, glutelin, lauric, and valeric acids, and starch than the control, and the antioxidant properties of the treated rice were weakened. CeO_2-NPs changed the content of amino acids, fatty acids, nonreducing sugars, and phenolics in plants.

2.11 Factors Affecting NP Phytotoxicity

In combination with recent studies and our own knowledge, we have concluded that the main factors influencing the effects of NPs on plants are the characteristics of the nanomaterials themselves (concentration, size, category, and stability), the plant seed (size and species), the plant growth medium, the plant growth stage, and the NP coating material. We will discuss the influence of various factors below.

The treatment of tomato with Ag-NPs resulted in a reduction in biomass and root length. TiO_2-NPs could significantly improve the germination rate of seeds; however, bulk TiO_2 can have inhibitory effects on seed germination. Because of the high reactivity of small materials, NPs are usually more toxic than larger-sized materials. However, other studies have shown that phytotoxicity increases with particle size. For example, all Ag-NP treatment groups had no effect on the growth of castor, but in the treatments where the Ag was used in bulk form, inhibition was observed. On the fifth day of NP exposure, synthesized silver NP treatments had higher Ag accumulation than biosynthesized silver NP (B-Ag-NP) treatments (10 and 100 mg/L) in water hyacinth; at the high concentration (100 mg/L), B-Ag-NPs improved plant growth.

To some extent, growth and antioxidant defense responses differ from species to species. Some studies have speculated that the differences in the toxicity of NPs might be related to the differences in seed size, the single leaf, and the xylem structure of double-leaf plants. We have mentioned the response of different plant seeds to the same nanomaterial in the previous section. The responses of different species to the same NPs must also be discussed. Three rice varieties [high (HA), medium (MA), and low (LA) amylase] were cultivated in soil amended with CeO_2-NPs; a comparison between the treatments showed that, relative to the control, the Ce content in HA was not significantly different. Treatment with CeO_2-NPs enhanced the Ce content in MA and LA treatments. Transgenic seeds are likely to have different responses when compared with conventional seeds. When exposed to ZnO-NPs, leaf surface area has been reported to reduce in *Oryza sativa* cv. Zhu Liang You 06 when compared with cultivar Qian You No. 1 under control conditions for both cultivars (Yang et al. 2017).

NPs must be studied in the specific environments in which they occur, as their characteristics differ widely depending on the environmental conditions. When the biomass of radishes was compared in silty loam (~2.21% soil organic matter) and loamy sand (~11.87% soil organic matter), the former had significantly higher root biomass, and the presence of 1000 mg/kg CeO_2-NPs made no substantial difference. The Ce accumulation in fine roots and in the storage, roots showed that growth in loamy sand was higher than that in silty loam. Kidney bean plants exposed to CeO_2-NPs for 52 days in a low organic matter soil usually possessed a higher Ce concentration than the same tissues collected from an organic matter-enriched soil. Likewise, different levels of phytotoxicity caused by CeO_2-NPs were found in lettuce seedlings incubated with three kinds of media, including agar, potting soil, and sand. Ag-NP toxicity decreased with increasing clay content and pH but seemed unaffected by the organic carbon content of the soil.

Photosynthetic parameters changed throughout the whole growth period in cucumber; seedling leaf size decreased when compared with the control when treated with 200 mg/L of CeO_2-NPs and CuO-NPs, but when mature leaves were measured, there were no significant differences among the treatments. At 15 days, the CAT and APX activities in the roots showed a sharp decrease when compared with the treatments at day 7, except for the control treatment. Total soluble protein content in roots showed the opposite results, except for the 500 mg/L CeO_2-NP treatment.

2.11.1 NP-Surface Coating and Phytotoxicity

Coating NP not only alters its function but also changes its effects on soil-plant systems and solubility. The suspension of Ag-NPs led to an increased germination rate for five of the 11 species, whereas exposure to coated Ag-NPs (GA-Ag-NPs) only contributed to the germination rate for one species. This was because of the presence of a co-ligand in the soil, which caused a preferential reduction in the bioavailability and toxicity of Ag^+. GA-Ag-NPs had a longer duration of toxicity than $AgNO_3$. No significant differences in the biomass production rate (fresh weight) of rice were found among plants treated with Fe-Co-filled CNTs compared to the non-stressed plants; MWCNTs and Fe-CNTs significantly decreased the biomass production rate at 30 and 50 mg/L.

2.11.2 NP Induced Toxicity/ De-toxicity in Plants

Regarding the effects of NPs on plants, recent reports have revealed ambiguous results. Many studies found that NPs had a positive impact, while others found that they had a negative impact. Plant species, NP properties (size, shape, types, structure, defects, and surface coating), and culture media might be giving the divergent results.

2.11.3 N-Induced Phytotoxicity in Plants

Generally, reactive oxygen species (ROS) contain both free radicals such as hydroxyl radical and superoxide radical and non-radical molecules such as singlet oxygen ($^1O_2$) and H_2O_2. ROS is the product of aerobic metabolism in an ordinary plant, which acts as signaling molecules, whereas excess ROS would cause adverse effects called oxidative stress, which emerges when the ROS level exceeds the defense mechanisms and is able to pose a threat to cells by inducing DNA damage, protein oxidation, electrolyte leakage, lipid peroxidation, and membrane damage, finally causing cell death. Published studies on metal and metal-based NP phytotoxicity suggest that NPs can induce oxidative stress in many plant species. The particle-dependent ROS formation and lipid peroxidation on the surface of cellular membranes occurred to clarify ZnO-NP phytotoxicity to ryegrass. Similarly, the phytotoxicity of Ag-NPs was caused by ROS generation, based on a high correlation between the decrease in viable cells and ROS generation at concentrations of Ag-NPs up to 10 mg/mL over a week of treatment.

Recently, the ROS-sensitive dye DAB (3,3′-diaminobenzidine) was used to find the accumulation of H_2O_2 in the plant roots treated with CeO_2 and La_2O_3-NPs. Results clearly showed that the insoluble, deep brown product was produced under the treatment of NPs and could be easily visualized by human eyes. Similarly, the treatment with 800 mg/kg CeO_2-NPs caused the accumulation of H_2O_2 concentration at 35 μM, which was almost 10 times higher than that of the control, whereas experimental results followed even though oxidative stress occurred, no lipid peroxidation occurred, and it did not induce ion leakage in either roots or shoots, which means that membrane integrity was intact. Because ROS H_2O_2 could be converted into more toxic •OH, which cannot be detoxified by any known enzymatic system, and because there are no relevant reports on •OH determination in plants caused by metal-based NPs exposure, cellular damage was unavoidable. Among all the ROS, the most reactive is •OH, which has a single unpaired electron. Therefore, it could interact with all biological molecules and cause subsequent cellular damages such as lipid peroxidation, protein damage, membrane destruction, and ultimately premature cell death.

When the ROS level exceeded the threshold, not only was normal cellular functioning harmed, but oxidative stress was exacerbated by the production of lipid-derived radicals, which could react with and damage proteins and DNA. Lipid peroxidation is the most damaging process in every living organism, and membrane damage usually acts as an indicator of lipid damage under all kinds of stresses. Malondialdehyde (MDA), one of the final products of the peroxidation of unsaturated fatty acids in phospholipids, causes cell membrane damage. It has been demonstrated that in plants exposed to various abiotic stresses, such as metal-based NPs, the lipid peroxidation occurring due to ROS generation is enhanced.

Generally, oxidative stress caused by products and/or ROS can induce covalent modification of a protein, called protein oxidation. Excess ROS generation may induce the modification of site-specific amino acids, fragmentation of peptide chains, aggregation of cross-linked reaction products, altered electric charge, and increased susceptibility of proteins to proteolysis. Oxidative stress-damaged tissues usually contain an enhanced concentration of carbonylated proteins, which are widely used as a marker of protein oxidation. Sulphur and thiol groups containing amino acids are highly vulnerable sites to attack by ROS. An H atom from cysteine residues can be abstracted by activated oxygen to form a thiyl radical, which will be cross-linked to another thiyl radical to form a disulphide bridge. In a recent study, the phytotoxicity and genotoxicity of Ag-NPs on germinating wheat seedlings were investigated. It was found that 10 mg/L Ag-NPs caused the alteration of all sorts of proteins, which are vastly related to cell metabolism. Likewise, exposure of *Oryza sativa* L. to Ag-NPs led to the identification of 28 responsive proteins, for example, decrement or increase in abundance, which were involved in various important physiological and biochemical processes, for example, Ca^{2+} regulation and signaling, oxidative stress tolerance, cell wall and RNA/DNA/protein direct damage, transcription, and protein degradation, ultimately leading to cell division and apoptosis. The phytotoxicity of Al_2O_3, ZnO, and Ag-NPs in soybean seedlings was studied at the proteome level; 16 common proteins in soybean leaves were significantly changed, which were predominantly associated with protein degradation and photosystem.

ROS can cause oxidative damage to nuclear, chloroplast, and mitochondrial DNA. The DNA molecule is a cell's genetic material, and any damage to the DNA would cause changes in the encoded proteins, which may result in malfunctions and/or complete inactivation of the encoded proteins. Enhanced DNA degradation has been observed in plants exposed to different kinds of environmental stresses, such as metal-based NPs and carbon nanomaterials. High concentrations of CeO_2-NPs had negative effects on the DNA of *G. max*. Similarly, *Allium cepa* roots suffered various negative effects, such as total chromosomal aberrations and an increased mitotic index, when exposed to bismuth oxide nanoparticles. Several studies have also investigated the potential genotoxic effects of TiO_2-NPs on various species of plants. A study on *Nicotiana tabacum* and *A. cepa* has found that genotoxic effects occur with exposure to different concentrations of TiO_2-NPs (100 nm).

Furthermore, damages to the chromosomes and micronuclei of *A. cepa* were observed, and DNA shearing and fragmentation were also found, causing the reduction in root growth. However, another study on castor bean (*Ricinus communis*) seeds under Ag-NPs exposure reported increased ROS production and associated antioxidant defense mechanisms, including promoting peroxidase (POD) and superoxide dismutase enzyme (SOD) activity, which caused an increase in phenolic acids. Phenols are synthesized in plants for defense against pathogens; therefore, ROS generation in this case may not be viewed as entirely negative as ROS also enhances root elongation up to a certain concentration.

The generation of ROS, which is induced by NPs directly or indirectly, plays a critical role in phytotoxicity mechanisms. The production of ROS is based on the physicochemical properties of NPs as well as the test species. Various determinants, such as size and shape, solubility and particle dissolution, metal ions released from metal and metal oxide NPs, biotransformation of NPs, light, and so forth, may cause ROS generation and phytotoxicity. The toxicity of three types of CeO_2-NPs to different kinds of Lactuca genus plants compared to the control, 7 nm CeO_2 caused a significant increase in the MDA level, which means the membrane damage happened in the root cells. However, there was no obvious discrepancy in the MDA level between the control, 25 nm CeO_2, and its bulk counterpart treatment. Other reports have studied the effects of the size and shape of NPs on various plant species as well. In addition, plant responses to CeO_2 exposure varied with particle sizes and the growth stages of plants.

Compared to high-solubility NPs such as ZnO-NPs and Ag-NPs, CeO_2-NPs are generally regarded as stable regardless of the biological or environmental systems involved and have always been used as a model material to study the phytotoxicity mechanisms of NPs. Recent studies have also demonstrated that CeO_2-NPs can be bio-transformed into $CePO_4$ and $Ce(CH_3COO)_3$ in a variety of plant species. The release of Ce^{3+} ions from NPs was confirmed to play a critical role in the phytotoxicity of CeO_2-NPs to *Lactuca* plants in agar media, aqueous suspensions, and potting soil. The research groups discovered that the CeO_2-NPs transformation occurred on the root surface of cucumber rather than in plant tissues, and they concluded that root exudates at the nano-bio interface play a critical role in the biotransformation of CeO_2-NPs. Furthermore, phosphates play a vital role in the transformation process of CeO_2-NPs in plants. It is well known that phosphates widely exist in the environment and are used as a basic component of many culture media for toxicity testing. A recent study has also confirmed that the phytotoxicity to romaine lettuce of CeO_2-NPs was determined by phosphates in a sand culture.

The phytotoxicity of ZnO-NPs on Lolium perenne (ryegrass) revealed that concentrated ZnO-NPs in the rhizosphere solution and absorbed on the root surface could affect plant growth. Results also revealed that the concentration of released Zn^{2+} ions was far below the toxic threshold for the plant, and, herein, the phytotoxicity effect cannot be attributed to ion dissolution. Toxicity by dissolution of meta-based NPs was also excluded from the main mechanisms in some other studies. Recently, a study was conducted on the toxicogenomic effects of CuO-NPs on *Arabidopsis thaliana* by using microarray analysis. The results also agreed with previous opinions that the released Cu^{2+} ions from CuO-NPs contributed to partial toxicity during CuO-NPs exposure. Ag-NPs were found to be more toxic to *A. thaliana* roots than Ag+ ions, despite having no effect on seed germination. Results also found that seedlings adsorbed very few amounts of Ag^+, which could not affect chloroplast structure compared to Ag-NPs. Ag-NPs could change the balance of oxidation and antioxidant systems and further affect the homeostasis of water and other small molecules within the plant body. A potential mechanism of phytotoxicity is that Ag-NPs adsorbed on the root surface, disrupt the structure of the thylakoid membrane, and decrease chlorophyll content, herein causing an inhibitory effect on plant growth. As for other NPs, the phytotoxicity of ZnO-NPs may also be attributed to their photocatalytic activity, which promotes the level of ROS production under irradiation with energy at/or above their band gap energy and induces phytotoxicity.

However, when the composition, concentration, size, morphology, and surface coating of NPs changed, the results of NPs would also change. Therefore, the phytotoxicity mechanism of metal-oxide NPs, for example, ZnO-NPs and Ag-NPs, may be derived from the NPs' physicochemical properties to impact seed germination rates, root elongation, and so forth, which need further research using proteomics and genomics-based investigations.

2.12 NP-Induced Detoxification in Plants

As mentioned above, certain environmental conditions, such as heavy metals, salt stress, nutrient deficiency, and diverse metal-based NP exposure, clearly affect the level of cellular ROS, for example, H_2O_2, $^1O_2$ and •OH. ROS is a signaling or damaging molecule that plays a vital role in the subtle equilibrium between the levels of ROS generation and scavenging. ROS have been involved in intercellular signaling cascades as second messengers, which regulate various plant responses in plant cells such as gravitropism, programmed cell death, stomatal closure, and so forth. They also regulate the activities of numerous components by signaling, including transcription factors, protein phosphatases, and protein kinases. However, excess ROS could cause oxidative stress in plants under NP exposure. To protect themselves against these toxic oxygen intermediates, plant cells and their organelles, such as mitochondria, peroxisomes, and chloroplasts, make use of antioxidant defense systems to scavenge excess ROS production. Antioxidant defense systems of plants contain both nonenzymatic antioxidants including thiols, glutathione (GSH), phenolics, ascorbate (AA) and so forth, and enzymatic components such as CAT, SOD, guaiacol peroxidase (GPOX), APX, glutathione reductase (GR), monodehydroascorbate reductase, dehydroascorbate reductase (DHAR), glutathione S-transferases (GST) and glutathione peroxidase (GPX).

2.12.1 Antioxidant Defense System: Non-enzymatic Agents

Thiols and AA are the most significant low-molecular-weight antioxidants. AA is a crucial antioxidant that can resist oxidative stress induced by the elevated level of ROS generation. Owing to the intrinsic ability to donate electrons of AA in diverse nonenzymatic and enzymatic reactions, it can scavenge $O\bullet-_2$, $O_2\bullet-$ and •OH and regenerate α-tocopherol from the tocopheroxyl radical to protect membranes directly. It also serves as the first line of defense against potentially harmful external oxidants. Shaw and Hossain et al. investigated the impact of CuO-NPs on rice seedlings. They found consistent increases in AA activity to ensure a higher scavenging rate of H_2O_2 in seedling leaves when exposed to less than 1.0 and 1.5 mM CuO-NPs. Similarly, it was observed that the 500 mg/L CeO_2-NPs treatment altered levels of AA and free thiols, causing enhanced membrane oxidative damage and photosynthetic stress in the shoots.

GSH, one of the fundamental metabolites (a nonprotein thiol with a low molecular weight) in plants, plays a key role in intercellular antioxidant defense against ROS-induced oxidative stress. In plant tissues, GSH is commonly found in all cell compartments, including chloroplasts, mitochondria, vacuoles, and the cytosol, as well as in peroxisomes and the endoplasmic reticulum. To preserve the normal state of cells, GSH plays an essential role in coping with the oxidative damage induced by ROS. As an antioxidant, GSH acts as a proton donor in the presence of organic free radicals and/or ROS, scavenging ROS and reducing them to a disulphide form, oxidized glutathione (GSSG). Additionally, via the AA–GSH cycle, GSH takes part in the regeneration of another potential water-soluble antioxidant product, such as AA. Although some researchers have begun to evaluate the levels of GSH in plants under metal and metal-oxide NP exposure, the reason for similar phytotoxicity caused by NPs is unclear. Recently, it was first reported that CeO_2 and In_2O_3-NPs caused sulphur assimilation gene regulation and GSH biosynthesis in *Arabidopsis*. GSH and/or GSSG quantification could be regarded as an alternative method to evaluate whether the biosynthesis of GSH plays an important role in the detoxification process under NP exposure. Likewise, enhancing the GSSG level in plants was induced by other NP exposure, whereas there is no direct connection between the enhanced level of GSSG production and the decrease in GSH, which converts H_2O_2 into H_2O.

Due to their antioxidant properties, phenolic compounds can scavenge ROS species, chelate transition metal ions, and inhibit lipid peroxidation by trapping the lipid alkoxyl ($-OCH_3$) polyphenols. Carotenoids are the member of lipophilic antioxidants and can detoxify many forms of ROS. Tocopherols, containing tocopherols (α, β, γ, and δ), represent a class of lipophilic antioxidants that participate in the scavenging of lipid peroxidation radicals, oxygen-free radicals and $^1O_2$. By chemically reacting with O_2 and physically quenching in chloroplasts, tocopherols can protect lipids and other membrane components, herein protecting the membrane structure and PS II. However, the investigation of phenolic compounds, carotenoids, and tocopherols in response to metal-based NPs exposure is very scarce, and the role of the above compounds against NPs-induced stress is also unclear. Understanding the response of the nonenzymatic defense system in plants under the exposure of NPs is very vital in accurately assessing the potential risks of NPs to plants soon.

2.12.2 Enzymatic Antioxidants and Antioxidant Defense System

SOD, the most effective intracellular metalloenzyme, plays a fundamental role in the antioxidant defense system against oxidative stress by catalyzing highly toxic ROS into less toxic H_2O_2 and O_2. It contains three isozymes, including iron SOD (Fe-SOD), manganese SOD (Mn-SOD), and copper/zinc SOD (Cu/Zn-SOD), which are ubiquitous not only in aerobic organisms but also in subcellular compartments to prevent against the toxic effects of enhanced levels of ROS production. Several reports have recorded that the promotion of SOD is related to the increase in tolerance of plants against environmental stress, such as NP toxicity, which means that SOD could act as an indirect selection criterion for researching oxidative stress. While working on tomato, significantly promoted levels of SOD were observed under $0 < 1000$ mg/L NiO-NPs exposure, whereas the decline of SOD activity revealed that the antioxidant defense system was compromised. It was recorded that the SOD activity in rice was elevated under 250 mg/L CeO_2-NPs exposure, whereas at a higher concentration treatment there was no obvious change in enzyme activity. Similar results were observed using 1 and 100 mg/L Al_2O_3-NPs exposed on onion root. Once H_2O_2 is overproduced either from alternative abiotic stress or from SOD defense system induction, to avoid oxidative stress such as lipid peroxidation, CAT, APX, and oxidized GSH contain the main antioxidant enzyme pathway that would be activated to detoxify H_2O_2.

CAT, among all the antioxidant enzymes, is the enzyme that was first discovered and characterized and plays an indispensable role in ROS detoxification under stress. It is widely found in tetrameric heme-containing enzymes and converts H_2O_2 into O_2 and H_2O directly. APX, regarded as the most significant ROS scavenger to protect higher plants, avoids oxidative stress, and plays a central role in regulating the AA–GSH cycle. Compared to CAT and POD, APX has a higher affinity for H_2O_2. APX can convert the low toxic H_2O_2 to nontoxic H_2O by AA oxidation into dehydroascorbate and MDA. GR takes part in both enzymatic and nonenzymatic oxidation–reduction cycles as an antioxidant. It is an NADPH-dependent enzyme that catalyzes the oxidation of GSH to GSSG, which can maintain a high ratio of GSH/GSSG in the cells. There are two steps involved in the catalytic mechanism. Originally, to achieve a cysteine and a thiolate anion, on interaction with NADPH, flavin is oxidized and a disulphide bridge is reduced. Then, the reduction of GSSG is involved in thiol-disulphide interchange reactions. GPX, a catalyzer, uses GSH to reduce lipid hydroperoxides and organic hydroperoxides, thus exhibiting a positive effect in plants against environmental stress. GR, APX, and DHAR are involved in the AA-GSH that can control the excess levels of ROS or the oxidative state in plants. It was also found that APX can reduce the level of H_2O_2, which was formed by SOD converting it to H_2O. Owing to DHAR's existence, regenerated AA was used by APX for the diminution of H_2O_2. Likewise, various antioxidant enzymes can be activated by numerous NPs; for example, Fe_3O_4-NPs, Co_3O_4 NPs, and CeO_2-NPs induced CAT; Au-NPs, MnO_2-NPs, Fe_3O_4-NPs, CuO-NPs, Co_3O_4-NPs and CeO_2-NPs induced GPX and fullerene; and Pt-NPs and CeO_2-NPs induced SOD. While working on *Phaseolus vulgaris*, it was noted that phytotoxicity in plants caused by CeO_2-NP exposure disturbed the antioxidant defense system. Several studies have demonstrated the activities of CAT and APX where plant species are exposed to NPs. In CeO_2-NPs-induced ROS generation, the CAT activity was reduced significantly in roots, suggesting that a detoxification pathway may be involved in the AA–GSH or GPX cycle. Compared to the control, the effects of TiO_2-NPs were recorded in cucumber, and it was found that NPs can induce a significant CAT activity ranging from 250 to 750 mg/L, whereas no obvious difference in APX activity was found, except a marked decrease at 500 mg/L. A dose-dependent response of SOD and CAT activity was employed in tomatoes under NiO-NP treatment. The defense mechanisms of *A. thaliana* found that the activities of CAT, SOD, POD, and APX were significantly ameliorated under CeO_2-NPs exposure, while elevations were only found for POD and SOD under In_2O_3-NPs exposure. Similarly, an article on the effects of Nd_2O_3-NPs and their bulk counterparts in pumpkin found that not only pumpkin growth was inhibited, but also the K, Ca, Mg, and S levels in pumpkin tissues were decreased under 100 mg/L Nd_2O_3 NPs and bulk exposure. They also reported that the enzymatic activities of SOD and POD were elevated, whereas those of CAT and APX were reduced, which could be due to other pathways scavenging H_2O_2 or their biosynthesis being inhibited by overproduced H_2O_2. This evidence indicated that the antioxidant enzyme response mechanism might be based on the NP concentration, plant species, media, and NP types. NMs can cause a great number of adverse effects on plants at both the cellular (cell membrane damage, chromosomal aberration, and chlorophyll synthesis disruption) and physiological (biomass decrease, root length inhibition, and so on) levels. Recent researchers are devoting their efforts to identify the mechanisms of plant defense against NMs-induced oxidative stress (Yang et al. 2017).

2.13 NPs, Plant Roots and Soil Systems

Metal-based NPs exhibit properties that allow them to spread through the environment, especially in the soil. These NPs affect the soil properties, consisting of pH, organic matter content, alkalinity, and water content; root exudates; microbial communities; and microbial activities. Consequent upon these interactions, homo-aggregations, hetero-aggregations, and dissolutions give rise to ionic species and sorption into biological systems. NP-concentration is demonstrated to influence their incorporation and accumulation in plants. One study reported that the mobility of Al-NPs in soil was lower than 10 m. Other reports noted that root uptake took place only during the first stages of NP mobility in the soil. It was also demonstrated that the mechanisms involved in NP movement in soil are sequestration and adsorption.

2.13.1 NPs in Roots

Several studies were carried out to explore the mechanisms of NP uptake as well as the barrier mechanisms in plants when they encounter metal NPs. Plant roots, which have rough surfaces due to the presence of root hairs, produce mucilage and different small molecules, including organic acids. Negatively charged root surfaces attract the positively charged NPs, as demonstrated in the translocation of CeO_2-NPs or neutral moieties compared to those with positive charges, which are more likely to remain adsorbed. While going through the processes of uptake and transport, NPs must pass through a series of physiological root barriers from the surface to the xylem vessels, including the root surface cuticle, epidermis, cortex, and endodermis, and are finally transported through the shoots via the xylem. To clarify whether plants uptake the nanoparticulate species directly or via dissolution before absorption, in this context, Ce adsorption was investigated in *Arabidopsis* plants, indicating that Ce was present at significantly higher levels in roots and shoots exposed to CeO_2-NPs as compared to those plants treated with bulk CeO_2. The results indicated that CeO_2-NPs were directly taken up by the roots. The uptake of CuO-NPs, bulk CuO, and ionic copper ($CuCl_2$) by *Capsicum annum* L. demonstrated that Cu levels in the roots of plants exposed to ionic Cu were higher as compared to the rest of the treatments. It was reported that ionic Cu reduced gas exchange parameters, evapotranspiration, stomatal conductance, and photosynthesis by an average of 41%, 59%, and 38%, respectively, as compared to the same Cu concentrations supplied in the form of NPs and bulk materials. Plants with diversified physical and chemical structures in roots, shoots, leaves, seeds, and fruits might show diverse mechanisms to face the presence of NMs. Factors like environmental conditions, the maturity of plants, and the physicochemical characteristics of soil play important roles in the above-mentioned phenomena. Numerous studies have confirmed that the symplastic route is the dominant pathway for the transport of NPs within plants, facilitated by important carrier proteins, aquaporins, and interconnected channels as well as the endocytosis process. Particulate forms of mineral nutrients are immobilized and mobilized via the xylem and phloem. Both transport systems involving the plasma membrane (apoplast) and plasmodesmal (symplastic) play an important role in the internalization, along with water, which is the carrier not only for nutrients but also for NPs. Water present in soil is absorbed by root structures involving primary roots, root cap, lateral roots, and root hairs and is radially mobilized through the xylem trachea. The movement of NPs is determined by the nature of these NMs and by plant biology. The water movement through the root apoplast is driven by pressure and an osmotic gradient that suggests that NPs travel through the epidermis, cortex, and finally the entire plant. After this first root exposure, two mechanisms occur: passive diffusion and facilitated transport.

The transport of active ingredients involved in plants comprises two types, namely, passive diffusion and the facilitated one described below.

2.13.2 Passive Diffusion

The passage of water across the cell wall is mediated by passive transport. Also, the passive diffusion of NPs is usually associated with the transport of low molecular-weight substances and endogenous metabolites across cell membranes. This pathway does not require any energy and allows the free movement of molecules from higher to lower concentrations. In experiments performed in this context, the presence of CeO_2 was demonstrated in the roots of corn plants grown in soil amended with these NPs. It was also revealed that alginic acid coating increased the transport of CeO_2-NPs to the root cortex. Similarly, citrate capping was found to increase the absorption and biotransformation of $Ni(OH)_2$ in hydroponically grown mesquite plants due to the presence of the citrate group. In addition, it was suggested that the biotransformation of citrate-coated NPs might have been initiated in the roots. Carrots exposed to TiO_2-NPs coated with hydrophobic and hydrophilic moieties in a recent hydroponics study demonstrated that differences in surface coating did not alter NP distribution through the plants, despite severe damage to carrot biomass. NP uptake was affected by their small size, and they entered the root system by breaking the roots' surface. The effect of CeO_2-NPs and Cd^{2+} on their uptake and accumulation by soybean seedlings grown in soil was reflected in the presence of increased Cd concentration in soybean roots and leaves, and thus the root barrier did not prevent Cd-uptake. These results also confirmed the presence of passive diffusion through plant cell walls and membranes. It is known that plant roots contain many ion and protein transporters in their membranes, and those might be acting concomitantly during the transport of other metallic ions. The presence of both species increased the production of root exudates, which may influence the dissolution of NPs, resulting in increased Ce uptake.

Another study found that Au-NPs (>20 nm) could be adsorbed onto the root surface and enter root cells via pores ranging in size from 5 to 20 nm. A portion of Au-NPs was also taken up by the plant, while other molecules remained adsorbed onto the root surface. The tomato roots showed a higher concentration of Au compared to rice roots, without this difference being explained.

2.13.3 Facilitated Transport

This belongs to the category of molecular or ionic transport across a biological membrane via specific transmembrane integral proteins. About 13 mineral nutrients, consisting of Fe, B, Cu, Mn, Mo, Si, Zn, Ni, Cl, Se, Na, Al, and Co, and required by plants in trace amounts, are transported into the plant with the aid of specific transport proteins (transporters/channels) present in plant roots. Despite the huge amount of information regarding mineral elements, there is still a dearth of information about the facilitation of the transport of NPs into plant roots. The uptake of Zn by *Phaseolus vulgaris* L treated with ZnO nanoparticles (40 and 300 nm) and $ZnSO_4$ (100 and 1000 mg Zn/L) for 48 h was described in a study of the physiological effects of these treatments. The results so far observed indicate that 40 nm ZnO-NPs are more easily dissolved by the roots than those with a size of 300 nm. Accordingly, root-to-shoot Zn-translocation presented a decreasing gradient of concentration and rate of absorption. The Zn movement was found to occur through the axial xylem system. Also, transcriptional analysis of genes encoding for metal carriers indicated a higher expression of tonoplastic-localized transporters regulating the content and volume of vacuoles in plant cells and are integrated by three major components: two proton pumps; the vacuolar H^+-ATPase (V-ATPase) and H^+-pyrophosphatase (V-PPase); and aquaporins. Tonoplast transporters include metal ion transporters like vacuolar Ca^{2+}-ATPase, vacuolar Ca^{2+}/H^+ antiporter, vacuolar Na^+/H^+ antiporter and Zn transporter. No specific information is there to determine whether these transporters are also involved in NPs-movement. NPs interaction with plants may involve adverse effects. An experiment involving 55 days exposure of *Helianthus annuus* L. plants to γ-Fe_2O_3-NPs concentrations of 50 and 100 mg/L in hydroponics revealed the blockage of the IRT-1 membrane protein transporters in the roots. In addition, a change in the spatial distribution of these proteins and the modification of complex systems, such as the water transport system, were also observed.

2.13.4 NPs in Stems and Leaves

NPs are reported to be transported through xylem and phloem in stems and leaves. Xylem and phloem are present throughout the plant system. In general, the xylem has two types of conductive cells: tracheids and vessel elements. Both cell types are connected to their neighbours via pits, which are small openings covered by a porous pit membrane. In addition to pits on their lateral walls, vessel elements are connected to each other end-to-end by perforation plates. Whereas the pit membrane can act as a barrier to solute movement, perforation plates allow their relatively unimpeded transit. The relative abundance of each type of conductive cell in the axial direction varies between plant species. Pit membranes of xylem conductive cells are found even when they are porous, restricting the passage of air bubbles and pathogens. Their pore diameters are in the range of 20–50 nm and are able to restrict NP transport, which might explain the accumulation of NPs in some tracheary elements. In addition, the differences in radii of xylem vessels in different parts of the same plant might affect the transport of NPs from root to leaf. These porous structures can prevent the transport of NPs through the conductive systems by acting as sieves. For the majority of the plants, the average sieve plate pores range between 200 nm to 1.5 μm, which should allow the passage of most NPs, unless they are in aggregated form. In the xylem and phloem, NPs are transported along with water and sap, in which two important physiological factors may impact the transport of NPs, namely: sap composition and sap flow rate. Sap composition is expected to affect the aggregation, sedimentation, and dissolution of NPs. Some studies revealed that significant concentrations of K, Ca, Mg, Na, Cl, P, N, S, sugars, amino acids, organic acids, and proteins are present in the phloem, although their concentrations vary over a wide range. In addition, the xylem contains inorganic ions but no sugars or organic acids, and the phloem presents higher concentrations of inorganic species. Also, sap composition depends on the growth stage and location of xylem or phloem within the plant (leaf), which affect water flow and nutrient transport. The composition of the sap is known to be very complex. It is, therefore, necessary to understand the influence of pH, ionic strength, inorganic ions (including Cl^-, SO_4^{2-}, and NO_3^-), and high concentrations of organic molecules in the context of their fate, transport, and transformation in plants. For example, it is likely that high ionic strength and abundant divalent cations would cause NP-aggregation in the sap. High ionic strength can lead to fast aggregation of NPs, while organic macromolecules enhance NP stability and reduce sedimentation or deposition. In addition, specific ions and organic macromolecules can accelerate or inhibit NP-dissolution. For example, at a low Cl/Ag molar ratio, Ag-NP dissolution was inhibited, while a high Cl/Ag molar ratio stimulated Ag-NP dissolution. The presence of humic or fulvic acids inhibited the dissolution of Ag-NPs, while extracellular polymeric substances promoted the dissolution of Cu-NPs. Different inorganic species participate in NPs' transformation. For instance, after FeO-NPs dissolution, the presence of phosphate, nitrate, and sulphide leads to the formation of $FePO_4$, Fe_3O_4, and FeS, respectively. Surface modification with synthetic organic matter, including olefin-maleic acid copolymer, sodium carboxymethyl cellulose, and poly-acrylic acid, among others, enhanced the transport of NPs in porous media by minimizing NP aggregation and deposition. Studies conducted to understand the fate and transport of NPs in plants have not been conducted, so this field offers an opportunity to deepen this line of research.

The sap velocity in the phloem varies between 0.07 and 0.58 mm/s, while that of the xylem is an order of magnitude higher, i.e., 0.47–4.8 mm/s. In other words, even at the lowest velocity (0.07 mm/s), NPs may travel up to 6 m in 24 h, suggesting that once NPs reach the conductive system of a plant, they can travel rather fast unless impeded by sieve plates or pit membranes. The sap velocity in the phloem remains quite constant throughout the day, while the xylem sap velocity is greater during the daytime than at night; also, the xylem sap velocity reduces to zero when stomata are closed.

2.13.5 Interaction of NPs in Leaves

The structure and function of leaf tissues are very different from those of the roots. Also, the upward transport system (xylem) differs from the downward transport system (phloem) in both structure and function. NPs uptake by plant is affected by several factors related with the nature of the NP itself, and also with plant physiology and the potential interaction of NPs with the environment. NPs are able to penetrate the leaf surface through stomatal pores. For example, *Lactuca sativa* leaves showed evidence of NP internalization when they were sprayed with Ag and TiO_2-NPs for 7 days. *Cucumis sativus* was also able to uptake CeO_2-NPs through the leaves. Other plants that displayed similar results include *Vicia faba*, *Zea mays*, *Arabidopsis thaliana*, *Triticum aestivum*, and *Brassica napus*. It has been reported that *Zea mays* plants were able to concentrate 50 μg Ce/gram of leaf when this organ was sprayed with a total amount of 0.4 g CeO_2-NPs (37 nm). No sign of Ce translocation to the stem was observed. Cultivation of *Arabidopsis thaliana* seedlings on agar containing TiO_2-nanoconjugates (2.8 ± 1.4 nm) and by using electron and X-ray fluorescence microscopy reported the penetration of NPs into the epidermis and underlying palisade tissue, which suggested a contribution of the stomata and endocytotic vesicles to NP absorption. Watermelon plants, which contain large stomata and vessels, were sprayed with different NPs (Fe_2O_3, TiO_2, MgO, and ZnO). These NPs initially presented sizes between 27.3–46.7 nm. Their size considerably increased in the suspension and significantly decreased during spray treatment. Particles that did not exceed the diameter of 100 nm penetrated the foliar stomata and were translocated from leaves to stems and roots. Polystyrene, CeO_2, TiO_2, Fe_2O_3, MgO, ZnO, Zn, Mn, and Ag are among other NPs that have displayed ability to enter the leaf pores.

2.13.6 NP-Cuticle Uptake

Plant-NP interactions take place in three steps: NP deposition on a plant surface (on a leaf, root, or stem); NP penetration through the cuticle and epidermis; and transport and transformation within the plant. The epidermis contains one or more cell layers covered by either a thick upper surface or a thin lower surface of cuticle. This waxy, hydrophobic cuticle contains very small pores. Thus, it is likely that this barrier against NPs is bypassed when the cuticle is damaged. Once NPs pass through the cuticle and the epidermis, or simply through the stomata, they have relatively easy access to the downward vascular system and the phloem. It has also been found that epicuticular waxes present on the leaf surface impose high water nano-suspension repellence, leading to short NP residence times and, in consequence, affecting NP uptake. Cuticle preferably repels charged NPs. However, foliar application promotes the uptake of NPs by stomatal openings, which are typically ~100 nm in diameter in plant leaves.

2.13.7 NP-Uptake by Stomata

Stomata in plant leaves may occupy up to 5% of the total leaf surface area and are used by the plant to regulate water and gas exchange with the environment. Stomata sizes fall in the range of 10′s of microns. For example, they are 5 × 13 μm on spinach leaves, 5 × 7 μm on watermelon and tomato leaves, 21 × 13 μm on cucumber leaves, and 16 × 6 μm on lettuce leaves. It is important to note that the location and quantity of stomata are plant-dependent. Most plant species only have stomata on the lower side of their leaves, while a few have them on both sides. When present on both sides, eudicots tend to have more stomata on the abaxial leaf surface (about 1.4 times that of the adaxial surface), while monocots have similar numbers of stomata on both sides. Several studies have demonstrated the uptake of hydrophilic substances through stomatal openings. The morphological size of stomatal apertures is approximately 25 μm in length and 3–10 μm in width. Even when their response depends on CO_2 concentrations, moisture, temperature, and light intensity, stomata are large enough to allow NP penetration when open. However, due to their unique geometric construction and physiological function, determining the range of stomatal apertures through which NP penetration is possible, has proven difficult. Different studies have found NPs or their aggregates in leaf stomata and deep tissues in different plant species, including *Citrullus lanatus*, *Cucurbita pepo*, *Allium porrum*, *Lactuca sativa*, and *Arabidopsis thaliana* as observed under TEM, CLSM, or μXRF. The uptake of 43 nm NPs through the stomatal pathway was detected with CLSM, whereas no uptake of 1.1 μm particles was observed. The fluorescent NPs were detected in the leaf apoplast after they entered the substomatal cavern. Plant species with different leaf morphology, stomatal size, and densities are expected to have different capacities for foliar NP uptake. In a recent study using single-particle inductively coupled plasma-mass spectrometry (SP-ICP-MS), lettuce, collard greens, and kale leaves were exposed to CuO-NPs (20–100 nm) and the results indicated that most of the NPs and their aggregates were rinsed with water. The capacity of the NPs to remain attached to leaf tissues was dependent on their hydrophobicity and leaf surface roughness. Similarly, 2.8 ± 1.2 nm TiO_2-NPs mostly accumulated in some of the stomata on an *Arabidopsis* leaf after 24 h exposure. Although it is difficult to differentiate the contributions of epidermal penetration and stomatal uptake to total NP uptake in foliar exposure, it is important to consider differences in stomatal aperture size and stomata density. Foliar exposure of ryegrass to sub-micron atmospheric dust containing Pb resulted in a Pb shoot content of about 700.1 ± 27.5 mg/kg dry tissue, with 14% of Pb in its elemental state and over 49% of Pb combined with organic acids or cell walls. Similar experiments using lettuce indicated a shoot Pb content of 171.5 mg/kg dry tissue, with 81% of Pb in its elemental state and 19% of Pb combined with organic acids or cell walls. It is, thus, considered likely that the stomata uptake pathway was responsible for Pb-uptake until reaching final concentrations of 98 mg Pb/kg dry shoot tissue in rye grass and 138.51 mg/Pb kg dry shoot tissue in lettuce. Even though this Pb study did not include NPs, this information may be helpful when comparing the uptake mechanisms of NPs and sub-micron particles through stomata. Accordingly, it is very likely that the stomata serve as an efficient pathway for NP uptake, with high stomata density leading to higher NP uptake. Therefore, it is important to further study the effect of stomata density, location, and the environmental variables that affect the opening of these pores on NP-uptake.

2.14 NP-Internalization in Leaf

The structure of the mesophyll, the inner leaf tissue, may affect the transport of NPs after they penetrate the epidermis and before they enter the vascular system. As previously mentioned, on foliar application, NPs enter the plant system through the cuticle, epidermis, and stomata. In some cases, NPs accumulate in the epidermis and are transported to other parts of the plant, as was the case with CeO_2 and CuO-NPs sprayed on cucumber plants. Another study investigated how the size (3, 10, and 50 nm) and coating of gold nanoparticles (Au-NPs) influenced the process of NP transport within the plant system. The results of this study indicated that following wheat foliar exposure to Au-NPs suspensions (280 ng per plant), adhesion on the leaf surface increased as NP size decreased. In addition, when comparing the interactions of PVP-Au-NPs and citrate-Au-NPs, some differences were observed. After 2 weeks of foliar exposure, part of the citrate-Au-NPs remained outside the cuticle layer. However, the fraction of citrate-Au-NPs that entered the leaf was translocated to the plant vasculature. In contrast, for similar sizes, all of the PVP-Au-NPs crossed the cuticle layer after 2 weeks. However, they were transported through the mesophyll cells at a lower rate. The study also indicated that, as a consequence of PVP-Au-NPs accumulation in the leaf mesophyll, wheat photosynthesis was impaired. The functional groups present in citrate and PVP are very different. For example, a ring and a N are present in every PVP monomer. This is not the case for citrate. However, there is no known data about the differences in the interactions of these two groups with both plant metabolites and cell plants. This is a relevant gap that needs to be filled. With the missing information, better NP coating designs will be possible. Phyllosphere factors affecting foliar uptake of NPs by plants in the natural environment, different symbiotic and pathogenic microorganisms are present in the phyllosphere. These microorganisms produce extracellular polymeric substances (EPS) and other chemicals that serve as quorum-sensing signal molecules. These exudates may induce the dissolution of NPs and promote the uptake of metal ions by plant leaves. They may also form a mucous layer, which protects the leaf surface and inhibits NP uptake. An important aspect is that the level of development in plant organs varies in different life stages, potentially influencing foliar NP uptake. For example, undeveloped cuticles in newly growing leaves and flowers as well as the natural senescence of old leaf cuticles may increase the probability of NPs entering. Some common leaf diseases such as leaf mould, blight, chlorosis, and necrosis may destroy the protective leaf tissues, including the cuticle, epidermis, and even the mesophyll. Therefore, leaf barriers against NPs may be disabled when pest or disease damage occurs. In particular, NPs such as $Cu(OH)_2$, TiO_2, and Ag/SiO_2, which are used as NPs, are sprayed on diseased leaf surfaces. This process may increase the risk of NP foliar uptake and further transport to other plant tissues.

2.15 NPs in Flowers and Fruits

Studies of NPs' interactions with different plant parts have been performed on leafy vegetables including lettuce and spinach, stem vegetables like asparagus and celery, fruits including tomato, zucchini, and legumes, and grains such as barley, rice, and corn. However, information on NP transport, interaction, and effects on fruits and flowers is scarce.

2.15.1 In Flowers

Few studies have reported the effects of NPs in flowers. The application of CuO-NPs to *Arabidopsis* increased the Cu concentration in flower buds as compared to untreated plants. Furthermore, flowers of zucchini plants grown in soils containing CeO_2-NPs presented Ce concentrations 1.5–1.9 times higher as compared to controls. Foliar application of TiO_2-NPs at common sunflower improved the stem elongation, flower-bud formation stage and induced seed development.

These effects could be related to the ability of TiO_2-NPs to enhance light absorption, light energy transformation, protect the protoplast, play a role as photocatalyst, and, therefore, improve photosynthesis. In contrast, plants treated with ZnO-NPs showed a prolonged sunflower vegetative phase related to fast-moving NP assimilation. ZnO and TiO_2-NPs induced changes in trichome distribution, width, and length on the surface of leaves collected at the flower-bud development stage. Even though NPs did not affect the quality of the oil and were not translocated to sunflower seeds. The negative effect of Ag-NPs on floral development could be transferred to the offspring in *A. thaliana*. Ag-NPs decreased petal and pollen viability and reduced pod production through the downregulation of genes related to the flowering process. However, Ag-NPs themselves were not transferred from the parent plant to the offspring plants. Studies of metallic NP translocation in flowers are in their initial stages. Since the flower is the reproductive organ of the plant, it is important to understand how it might be affected since plant propagation and survival may also be affected.

2.15.2 In Fruits

Experimental observations collected in this context often indicate that accumulation occurs mostly in roots rather than other tissues. Nevertheless, exposure to NPs or metallic oxides may produce an increase in the fruit's metal content. The addition of CuO-NPs to the soil medium where bell pepper plants were growing led to an increase in the amount of copper in the fruit, with no differences when bulk CuO materials were used. In addition, when CuO-NPs were applied at concentrations of

500 mg/kg, they inhibited zinc absorption in the fruit without presenting any visible signs of toxicity or productivity effects. Approaches via metabolomics for understanding the implications of NPs on the nutritional value of cultivated fruits have been published. It was noted that CuO-NPs altered cucumber's metabolic profiles by reducing myo-inositol, methionine, citric acid, ornithine, and citrulline content. At the same time, these NPs led to an improved concentration of sugars, amino acids, organic acids, and fatty acids that are beneficial for the quality of the fruit. On the other hand, studies conducted on tomato plants grown in pots supplemented with CeO_2 and Ce-NPs coated with citric acid (nCe + CA) confirmed that both NPs were capable of differentially modifying the nutritional quality of tomato. The nCe + CA treatment resulted in a decrease in dry weight and fruit sweetness that was associated with the presence of citric acid. An increase in B concentration was observed, which was probably due to the interaction between this element and the negative charge of the NPs. In addition, CeO_2-NPs caused a decrease in the absorption of B, Fe, Mn, and Ca due to these cations' repulsion. It is important to note that it was not possible to detect Ce content in the fruits.

Recently, the phytotoxic effects of NPs were found depending on their surface properties. Zn, Cu, and Ce can accumulate in the pulp of sweet potatoes grown in soil containing ZnO, CuO, and CeO_2-NPs as well as their corresponding metallic ions. The three metals accumulated in the epidermis and pulp of sweet potato tubers, with higher concentrations in the epidermal peel. Results did not differ when treatments contained either the NPs or their metallic ionic forms. This suggested that NPs were first dissolved, and the ions released before absorption by the potato tuber. Similar effects were observed in carrots grown in sand containing ZnO, CuO, and CeO_2-NPs. The results showed that the outer layer of the root (epidermis) served as a filter and limited the accumulation and distribution of metallic NPs in the carrot taproot flesh. A decrease in the absorption of some minerals was reported in the plants treated with NPs as compared to those treated with the ionic forms (Zn^{2+}, Cu^{2+}, or Ce^{4+}). In addition, NPs aggregation in the media may have occurred. In a previous study carried out in *Raphanus sativus* (L), the addition of CeO_2-NPs to the soil caused an increase in Ce accumulation and antioxidant capacity in the tubers without affecting nutrient concentration. It has also been shown that CeO_2-NPs affect the flavor and antioxidant capacity of cucumber. Antioxidant capacity was reduced because NPs affected the concentration of Mo, non-reducing sugars, phenolic compounds, and the balance of the protein profile. In the same way, Ti content in the secondary roots of carrot plants treated with different TiO_2-NPs was influenced by the surface characteristics of the NPs, mainly those with hydrophobic moieties. All treatments resulted in abnormal taproot splitting and decreased Zn accumulation. The micronutrient concentrations in the fruits decreased after NP exposure. Previous reports have indicated that ZnO-NPs caused a decrease in Cu and Mo concentrations in cucumber, although they led to an increase in the caloric value by modifying the starch and protein content.

The absorption of NPs by edible vegetables is still not fully understood at the cellular level. In a study in which romaine lettuce was exposed to CeO_2-NPs, it was discovered that the Ce accumulated in the sprouts was in the form of $Ce(CH_3COO)_3$, as previously reported, while CeO_2 and $CePO_4$ were detected in the roots. It has been suggested that the transformation of Ce in plants is due to the interactions between NPs and root exudates. Biotransformation caused a decrease in the translocation and toxicity of CeO_2-NPs, indicating that the release of Ce^{3+} from NPs was essential. However, translocation depends on the NP type. In another experiment, rutile and anatase were detected in cucumber fruits exposed to TiO_2-NPs. These findings demonstrated that both TiO_2 crystalline phases were transported from the soil to the edible plant part with no biotransformation. However, this behavior depends on specific characteristics (pH, organic matter, ionic strength, microbiota) of the soil used for the crop cycle since they affect NP solubility, bioavailability, and toxicity.

2.16 Analytical Techniques

Proper analytical techniques to study NPs' interaction with plants should be able to provide precise information on the presence of these materials within plant tissues as well as potential chemical modifications. Preferred techniques should include microscopy and spectroscopy. Extremely robust options include the use of synchrotron radiation (SR), such as synchrotron radiation X-ray fluorescence mapping (SR-XFM) and spatially resolved X-ray absorption spectroscopy (XAS), among others. These techniques have provided important information for a better understanding of NM's behaviour and fate. Another technique is single particle inductively coupled plasma mass spectrometry (SP-ICP-MS), capable of simultaneously detecting both particulate and dissolved forms of the analyte without complicated sample preparation steps. Thus, selecting the proper analytical methods is crucial to formulating robust conclusions.

The impact of NPs on plants, which represent a central component of ecosystems and the world's food supply, is of prime importance. In this context, NPs in the form of nanofertilizers and nanopesticides represent an option to increase crop productivity. NPs interact with plants via different uptake, translocation, and accumulation mechanisms that may involve modifications to plant metabolism. As a result, positive and negative effects may occur. Relevant gaps that need to be filled in order to better understand these effects are highlighted as follows. Metallic NPs present in the soil systems interact with soil media components. These interactions may modify the physicochemical and biological parameters of NPs, eventually resulting in: aggregation; dissolution; and biotransformation. These modifications will affect NPs' interaction with root cells. It has been proven that NPs migrate within the soil, and these velocities have been measured. The first gap is that no models have been provided that explain how soil characteristics influence the velocity of migration within the soil. In addition, the influence of biota has not been considered in the measurement of this parameter. At the root level, a significant amount of research has been performed, and important data have been published in the last few years. It has been proven that specific capping agents such as alginic and citric acid facilitate NP uptake. Carboxylic groups are able to attract cations and are involved in NP dissolution. However, it is unknown if they also interact with carrier proteins or membrane channels to aid in the uptake process. It is also known RT-1 membrane protein transporter is blocked in the presence of γ-Fe_2O_3-NPs, and several other proteins are modified. How these modifications occur is an important question to be answered. Furthermore, data have indicated that certain NPs with sizes smaller than cell wall pores are able to pass through and enter the root system. However, specific transport details are unknown. NPs transport from root to stem and leaf is carried out through the axial xylem system. In addition, it has been determined that several tonoplastic-localized transporters are overexpressed in the presence of NPs. These transporters affect the content and volume of vacuoles, which participate in NP storage. Even when the over-expression of these proteins has been proven, no direct evidence of NP-transporter interaction has been provided. This is another gap to be filled. During foliar NPs application, some particles enter the leaf system via stomata openings. Moreover, it has been proved that in some cases, NPs pass through cuticle pores. However, the specific mechanisms are unknown. Since cuticle is a waxy hydrophobic component, the NPs with hydrophilic capping would be repelled. However, this has to be proven. It has also been found that after foliar application, most NPs do not enter the plant, and a small fraction of them remain trapped in the epidermis. In addition, a limited number of particles that are able to penetrate the leaves are effectively transported to other plant parts. Thus, it is important to further study the effect of stomata density, stomata location, and the environmental variables that affect the opening of these pores on NP uptake, in addition to the mechanisms by which NPs are stored in the leaves without further transport. In order to solve potential problems related to the low NP availability during foliar application, leaf feeding or trunk injection could potentially deliver a relatively large number of NPs into the plant. This point also needs to be proven.

From a risk assessment perspective, it is crucial to assess the accumulation and transfer of NPs through various food chains and determine the fate of NPs within products for human consumption. Insufficient studies have been performed to determine whether these NPs may enter through the stem surface. Given that some NPs may be present in the atmosphere and be eventually deposited on the stem surface, it would be interesting to determine whether this might be a port of entrance, as well as the mechanisms involving potential transport to other parts of the plant system, including potential bio-transformations. Another relevant aspect involves understanding the selectivity of biological structures in plant cellular tissues that facilitate the internalization of NPs, especially metallic ones. In similar terms, it is crucial to determine the specificity of the proteins involved in the NM-uptake and how these proteins may adapt to specific characteristics of the NPs, such as size and charge. Also, since the physiology of plants is highly diverse, it is crucial to define how structural differences as well as plant growth stage may regulate the kinetics of NP uptake. From a more systemic point of view, it is necessary to understand how biotransformation is related to NP uptake and transport.

2.17 Chapter Summary

An effort has been made to describe the possible biomolecular interactions of the nanoparticulate material species studied in plants and crops. Taking into consideration the molecular transport via several feasible biomolecular pathways, it has been found possible to explore the idea of the translocations of the nanoparticulate material species in different parts of the plant, right from the roots to the leaves, flowers, and fruits of the plants involved. A brief introduction of the experimental observations gathered by several research groupsare noted to corroborate the theoretical possibilities and these information's are necessary to examine the related biomolecular interactions involving modified physicochemical properties due to nanoscaled morphologies via transport modified due to the modified electron states involved due to quantum size effect invariably present there. This kind of background information would make the readers more curious about the various useful and harmful effects of the interactions in general.

Further Reading

Gleiter H (1989) Nanocrystalline materials. Prog Mater Sci 33:223–315. https://doi.org/10.1016/0079-6425(89)90001-7

Henglein A (1989) Small-particle research: physicochemical properties of extremely small colloidal metal and semiconducting particles. Chem Rev 89:1861–1873. https://doi.org/10.1021/cr00098a010

Mout R, Moyano DF, Rana S et al (2012) Surface functionalization of nanoparticles for nanomedicine. Chem Soc Rev 41:2539–2544. https://doi.org/10.1039/C2CS15294K

Sanzari I, Leone A, Ambrosone A (2019) Nanotechnology in plant science: to make a long story short. Front Bioeng Biotechnol. https://doi.org/10.3389/fbioe.2019.00120

Scher JA, Elward JM, Chakraborty A (2016) Shape matters: effect of 1D, 2D, and 3D isovolumetric quantum confinement in semiconductor nanoparticles. J Phys Chem C 120:24999–25009. https://doi.org/10.1021/acs.jpcc.6b06728

Yang J, Cao W, Rui Y (2017) Interactions between nanoparticles and plants: phytotoxicity and defense mechanisms. J Plant Interactions 12:158–169. https://doi.org/10.1080/17429145.2017.1310944

Chapter 3
Nanomaterials and Phytonanobiotechnology

3.1 Introduction

Nanoparticulate materials of inorganic, organic, and biomolecular origins have extensively been explored for their numerous applications, including plant sciences. For obvious reasons, more examples of these applications are available in the field of nanobiomedicine, but the experience gained there has also been used to very effectively extend them to deal with similar situations in plant sciences, including modern crop management. Several descriptions of their uses in pharmaceutical and diagnostic applications are purposefully included in this work to highlight the various approaches being investigated in the context of NM-biomolecular interactions in general. Once the stable and reliable nature of the interactions explored with a set of NPs and small/macromolecular species is established, it finally becomes a part of a strategy that can be deployed elsewhere as well. Such interactions do not distinguish the origin of the species involved, but rather go by the availability of chemical affinity within the group of interactants and the nature of the surroundings.

3.2 Biosensor Applications

NMs have been used in fabricating biosensors for crop biotechnology, agriculture, and the food industry. Different categories of nanosensors have been tested in plants, including plasmonic, FRET-based, C-based electrochemical, nanowire, and antibody nanosensors. Initial reports have already confirmed the use of NMs as tools for detection and quantification of plant metabolic flux, pesticide residues in food, and bacteria, viral, and fungal pathogens. Onion membrane-based fluorometric optical sensor for detection of sucrose-based synthesis of invertase-nanogold clusters embedded in plant membranes was recently reported. In addition, SWNTs were

A. Nazeer et al., *Targeted Delivery of Nanopesticides and Nanofertilizers in Sustainable Agricultural Farming*, Nanotechnology in the Life Sciences,
https://doi.org/10.1007/978-3-031-41333-9_3

used in NIR-fluorescence monitoring of nitric oxide in *A. thaliana*. Polystyrene-NPs-conjugated FRET probes were also found useful in recognizing and quantifying the phytoalexins.

These NMs-based biosensors were found to enable rapid detection and precise quantification of fungi, bacteria, and viruses in plants. For example, fluorescent silica-NPs combined with antibodies were found useful in diagnosing *Xanthomonas axonopodis* pv. *vesicatoria*, which causes bacterial spot disease in Solanaceae plants. Recently, Au-NPs were proposed as DNA biochemical labels to detect *Pseudomonas syringae* in *A. thaliana* by differential pulse voltammetry on disposable screen-printed carbon electrodes. Similarly, fluorescently labeled-DNA oligonucleotide conjugated Au-NPs were used in the diagnosis of the phytoplasma associated with the flavescence dorée disease of grapevine. Smart nanosensors are becoming increasingly available for mycotoxin detection and a competitive antibody assay-based sensor was successfully introduced in the market to test the presence of ZEA, T-2/HT-2, DON, and FB1/FB2 mycotoxin residues in corn, wheat, oat, and barley (Wieszczycka et al. 2021).

Another tool for cellular imaging employed the MRI technique for imaging the soft tissue, in which some contrast improvement was needed to distinguish the normal tissue from the lesions by deploying a NP-based contrast agent for enhancing the MRI sensitivity. Currently, iron oxide NPs, iron nanocrystals, and iron-based core-shell nanostructures, due to their high magnetic saturation, stability in biological media, and low toxicity, have been extensively explored *in vitro* and *in vivo* as effective MRI contrast enhancers. Multifunctional hybrids, Fe-C core-shell NPs, were employed as superior T2-contrast agents. The iron-NP-core, in comparison to iron oxide-NPs, is less stable and biocompatible, thus its surface was coated with a carbon shell, causing no degradation of the magnetic properties. Iron and iron oxide-NP surfaces were also modified with silica, PEG, or various peptide shells to improve stability and shelf life. Hybrid core-shell NMs doped with gadolinium, Gd-NPs, as well as ultrasmall gadolinium-NPs, have been investigated as T1-weighted MRI contrast agents in chelated form or incorporated into nanostructures. Bimetallic Gd-Au core-shell NPs immobilized in biopolymers were found to improve Gd biocompatibility. Based on *in vivo* biodistribution and toxicity studies, the cellular internalization of core-shell NPs was found to provide detection via T1-weighted MRI.

A variety of INMs have been used to create nanosensors for the selective detection of waterborne microbes, with surfaces modified with molecular recognition entities such as peptides, antibodies, or aptamers to provide target selectivity. These strategies were used in fabricating optical, electrical, and magnetic sensor platforms. Optical nanosensors use fluorescence or localized surface plasmon resonance phenomena. Several QDs, up-conversion NPs, and metal NPs are still being explored as pathogen sensing platforms. In one such nanosensor, the electrochemical current and voltage characteristics were measured due to electron transfer between the NM's surface and the target analyte. Only highly conductive and electrochemically stable NMs were attached to the surfaces of these sensors (Wieszczycka et al. 2021).

In the case of magnetic nanosensors, the detection of pathogens employs interactions between biomolecule-functionalized magnetic NPs and the target analyte. The magnetic NPs allow isolation of the detected pathogens. The application of INPs in electronics, including printed electronics (PE), is a dynamic branch of nanotechnology and materials science that results in the preparation of printed and flexible electronic devices by printing with NM-based inks. A number of metallic NPs were investigated for PE fabrication, including gold, silver, copper, their alloys, or core-shell structures. Functional NPs could improve the ink's properties in terms of its viscosity, melting temperature, conductivity, and stability, which would finally reduce the overall production costs. INPs' higher electrical conductivity and better thermal and chemical stability make them suitable for many applications. Multifunctional INPs are used in energy conversion and storage, including nanostructured solar cells, electrodes for batteries, supercapacitors, or nanocomposites for hydrogen storage. Metal NPs, QDs, and up-conversion NPs might enhance the optical absorption and efficiency of solar cells due to surface plasmon resonance or light scattering phenomena. In dye-sensitized solar cells, porous INMs are used as counter electrode substrates. The high active surface of porous nanostructures ensures a higher number of catalytic sites, which are important for the reduction of oxidized charge mediators (Wieszczycka et al. 2021).

3.3 NMs in Plant Genetic Engineering

The cell wall barriers faced during the delivery of exogenous biomolecules in plant cells have been overcome to achieve plant genetic transformation, and consequently, different strategies based on *Agrobacterium* transformation or biolistic methods were used for DNA delivery in plant cells. The limitations of these approaches are based on limited host ranges and extensive plant damage, which frequently inhibit plant development.

Most of the studies for NM-based plant genetic engineering have been conducted in plant cell cultures. For example, SiC-mediated transformation was reported for successful DNA delivery in different crops of tobacco, maize, rice, soybean, and cotton. Although lagging the advancements made in animal systems, results reported recently in plants are proving that NMs may overcome the barrier of the cell wall in adult plants and reduce the drawbacks associated with current transgene delivery systems.

One such study confirmed that dsRNA from different plant viruses can be loaded onto non-toxic, degradable, layered double hydroxide (LDH) clay nanosheets. The dsRNAs and/or their RNA breakdown products protect against the Cauliflower mosaic virus in sprayed tobacco leaves, but they also confer systemic protection to newly emerged, unsprayed leaves on viral challenge 20 days after a single spray treatment in tobacco. This is a proof-of-concept for species-independent passive delivery of genetic material without transgene integration into plant cells for different biotechnological applications in plants.

A successful genetic transformation was reported in cotton plants via magnetic NPs complexed with β-glucuronidase reporter gene, which was infiltrated into cotton pollen grains by magnetic force, without compromising the pollen viability. Cotton transgenic plants were successfully generated via pollination with magnetofected pollen,and exogenous DNA was successfully integrated into the genome, effectively expressed, and stably inherited in the offspring obtained by selfing.

In another recent study, CNT-scaffolds were applied to external plant tissue by infusion to deliver linear and plasmid DNA, as well as siRNA, in *Nicotiana benthamiana*, *Eruca sativa*, *Triticum aestivum*, and *Gossypium hirsutum* leaves, and in *E. sativa* protoplasts, resulting in a strong transient green fluorescent protein (GFP) expression. Moreover, small interfering RNA (siRNA) was delivered to *N. benthamiana* plants expressing GFP, resulting in ~95% silencing. The first method for genome editing mediated by mesoporous silica-NPs was recently proposed, in which MSNs were used as carriers to deliver Cre recombinase in *Zea mays* immature embryos carrying loxP sites integrated into chromosomal DNA. The loxP recombined properly after the biolistic introduction of engineered MSNs in plant tissues, establishing a successful genome editing protocol (Sanzari et al. 2019).

Nanomaterials are currently being studied for their applications in monitoring and controlling plant functions. Integrating the ENMs with plants is leading to better advances towards addressing crucial challenges in plant genetic element delivery, biochemical sensing, and nutrient and pesticide delivery. The broad potential for engineering plants using ENMs with unique physical and chemical properties relies on circumventing plant barriers, including cell walls and membranes, and improving the targeting of specific tissues and organelles. The targeted delivery in non-plant eukaryotic cells could be obtained by surface functionalization of ENMs with guiding moieties. Although *in vivo* delivery efficiencies of NPs attached with biorecognition ligands to target cancer cells are not higher than 2%, localization of gold NPs coated with SV40 large T antigen in the nuclei of HeLa cell cultures *in vitro* has been reported to be up to 60%. Previous studies reported the non-targeted delivery of DNA and biochemicals into plant cells and organelles by MSNs through particle bombardment of leaf sections *in vitro* or by interfacing CNTs and their cargoes with isolated tobacco cells. Current approaches to improving *in vivo* NP-delivery efficiency to specific plant cells or organelles are based on modifying NP properties such as size and charge but do not reach high levels of subcellular localization specificity. For example, negatively charged cerium oxide NPs delivered into plant leaves have ~45% colocalization rates with chloroplasts. However, most chloroplasts do not contain NPs. The phenomenon of targeted delivery of NMs guided by biorecognition ligands like transit peptides have not been reported in plants to date because they cannot be directly translated from non-plant systems. Plant cells have a wall that acts as an additional barrier for NP translocation. NP-uptake across plant cell walls is limited by the size of NMs and cell wall pores. Despite conducting no systematic characterization of the permeability of plant cell walls to the NMs, it is expected to be dependent on plant species and NP properties. For instance, amphiphilic NPs (~40 nm) have been reported to translocate across leaf cells but not hydrophilic NPs of similar or larger sizes. Furthermore, strong leaf background fluorescence from chloroplast pigments impairs the easy tracking and colocalizing of the NMs in plants, thus requiring a specific design of fluorescent NMs for both targeted and traceable biochemical delivery in plants with a low fluorescent background.

Chloroplasts are important organelles for plant bioengineering because they are essential for plant photosynthesis, act as signalling organelles, and actively participate in metabolite synthesis. The chloroplasts are thus important for improving the crop growth, stress tolerance, biopharmaceutical production, and developing synthetic biology tools. Manipulation of chloroplast function through biomolecules or chemical delivery in plants is crucial for understanding the role of these plastids in plant biology and developing approaches for chloroplast bioengineering. However, chloroplast studies have been limited to the need for targeted delivery platforms for specifically engineering plant organelle function *in vivo*. Instead, chloroplast studies mainly rely on the generation of transgenic or mutant plants with altered development and function in a handful of model species amenable to genetic transformation. The design of novel ENM platforms was reported involving a peptide recognition motif to target the delivery of chemicals to chloroplasts in wild-type *Arabidopsis thaliana* (Col-0) plants. Using these nanoplatforms, a capability to specifically manipulate chloroplast function and redox status *in vivo* was reported (Santana et al. 2020).

In one of the studies of *in vivo* imaging, a novel ENM platform was designed and synthesized so that it could bypass the biological barriers in plants such as cell walls, membranes, and organelle envelopes for *in vivo* traceable and targeted delivery of biochemicals to chloroplasts using peptide recognition motifs in a hydrophilic QD (β-cyclodextrin peptide) to recognize organelle membranes, translocate across lipid bilayers, and emit optimal fluorescence while avoiding the leaf background. The QD intrinsic fluorescence and heavy element composition (cadmium and tellurium) allowed their specific imaging in plants by high spatial resolution confocal fluorescence microscopy and quantitative detection through advanced analytical tools including elemental analysis by ion coupled plasma mass spectrometry (ICP-MS).

Multifunctional fluorescent QDs were found as traceable chemical delivery platforms by forming inclusion complexes with chemicals like methyl viologen (MV) and ascorbic acid (Asc) on their surface through conjugated β-cyclodextrins. Tunable changes in chloroplast redox status were achieved specifically in this organelle by inducing or reducing superoxide anion production in the permitted delivery of MV and Asc. Manipulation of chloroplast redox status was associated with wide genetic and physiological responses in plants. A chloroplast targeting peptide, designed from Rubisco small subunit 1A (RbcS, genbank: OAP15425), was used to functionalize fluorescent QDs for targeting chloroplasts in the leaves of plants *in vivo*. This turned out to be the first experiment guiding NPs to a specific chloroplast by mimicking the biorecognition for protein precursor delivery. An alignment of the peptide amino acid sequence with RbcS-peptide analogues from multiple dicotyledonous plant species indicated a high degree of conservation in its composition and sequence across crop and model plants. The transit peptide (RbcS) enabled cytosolic recognition of the proteins for import into plastids by the chloroplast outer membrane. TOCs recognizing the N-binding domain of most pre-proteins destined for plastids acted in coordination with translocons at the inner membrane of chloroplasts (TICs) to allow import of pre-proteins into the chloroplast stroma. The ENM targeting sequence was truncated to the first 14 amino acids to minimize the increase in hydrodynamic diameter of QDs and improve the penetration through leaf cell wall pores. A short sequence containing GGC was added to the C-terminal of the peptide as a spacer. The terminal cysteine residues were further utilized as conjugation sites for binding with NHS-PEG$_4$-MAL (succinimidyl-[(N-maleimidopropionamido)-tetraethyleneglycol] ester)-linker onto the QDs. The NPs were highly conserved among dicots, containing functional biorecognition motifs that allowed translocation across the chloroplast double membranes (Santana et al. 2020).

QDs with β-cyclodextrin conjugated with targeting peptides (Chl-QD) permitted the targeted delivery of biochemical cargoes into chloroplasts. The β-CD molecular basket composed of seven cyclic oligosaccharides provided host-guest formation with ascorbic acid or methyl viologen and could form inclusion complexes with a broad range of biomolecules and chemicals including metabolic intermediates (β-carotenes), or herbicides (MCPA and norflurazon). At the terminal amine group located on the β-CD, a succinimidyl-dPEG-maleimide linker (NHS-PEG$_4$-MAL) was added to provide a selective conjugation site for cysteine residues located on the transit peptide. The QDs of 4.3 ± 0.2 nm size, measured by TEM, and the Chl-QD average hydrodynamic diameter of 24.5 ± 2.5 nm, measured by dynamic light scattering (DI water, pH 7), were the maximum sizes of NMs reported to translocate across leaf cell walls. The synthesized Chl-QD were negatively charged with a zeta potential of -28.4 ± 3.8 mV in DI water to facilitate translocation across plant lipid membranes. NPs with high zeta potential were reported to passively penetrate through chloroplast envelopes and plant cell membranes. Furthermore, QDs possess higher and stable fluorescence offering in vivo tracking within plant tissues and cellular compartments. The Chl-QD fluorescence peak was tuned to 540 nm to reduce interference with the leaf background and fluorescent dyes used in this study. The QDs exhibited a characteristic absorption peak at 465 nm in the UV-vis absorption spectrum. In Fourier transmittance infrared spectra (FTIR), significant characteristic bands for asymmetric glycosidic vibration (C-O-C) denoting β-cyclodextrin were detected at 1058 cm^{-1}, and bands typical of type I and II amide bonds at 1615 and 1515 cm^{-1} confirmed the successful conjugation of β-cyclodextrin and RbcS peptide onto QDs forming Chl-QD complex.

Precise control of NP location in plants is the most important aspect of using nanotechnology in plant biology, biotechnology, and nano-enabled agriculture. Using guiding peptides was noted to provide a molecular tool for targeted delivery of NMs to subcellular compartments with a precision not achieved earlier in plants. The difficulty of successful delivery was removed by controlling QD-size, zeta potential, and fluorescence emission range, allowing in vivo tracking with low interference from leaf background. Although Cd-based QDs were ideal for detection by multiple advanced analytical tools, including confocal fluorescence microscopy and elemental analysis by ICP-MS, chronic or high-level exposure to these NPs could be toxic to organisms depending on their surface properties. QDs within the non-toxic exposures were found useful for basic research at the interface of plants and nanotechnology, but applications in agriculture would require the use of environmentally friendly Cd-free QDs. However, a peptide recognition motif functionalized QD demonstrated the ability to engineer targeted NMs for improved colocalization within organelles and delivery of biochemicals *in vivo*. The high localization efficiency and specificity of NMs enabled by biorecognition surpass those of conventional methods based on NP-size and charge alone. The NP-mediated delivery of biochemicals such as MV and Asc to chloroplasts specifically allowed tuning of their redox status while minimizing ROS generation or scavenging in other subcellular compartments.

The targeted delivery of MV and Asc to chloroplasts exhibited ~79% efficacy, but further research on modifying the guiding peptide, including spacer amino acids, hydrophilicity, and charge, among other factors, is needed to enhance the delivery specificity to plant organelles while avoiding unwanted exposure to other plant cell compartments. For NPs coated with ligands larger than those used in this study (e.g., plasmid DNA and RNA), increasing the amino acid spacer length may be required. Tuning the hydrophobicity of the peptide recognition motif is also likely to affect the NP-translocations in plant cells. Models of plant cell and chloroplast uptake based on NP-size and charge alone using polymeric ligands do not explain why Chl-QD coated with biorecognition motifs with a lower zeta potential than that of MPA-QD exhibit higher colocalization rates with chloroplasts. This indicates that NP-plant interaction models should incorporate engineered biomolecule coatings and coronas for making accurate predictions of the distribution of NPs in plants. The biorecognition motif targeted delivery methods could be translated to a wide range of plant species that highlight the strategies to create nanobiotechnology tools that can bypass plant biological barriers for targeted delivery of biomolecules and chemical cargoes to chloroplasts for fundamental research in plant biology and more efficient delivery of agrochemicals to crops. The biorecognition approach of coating nanoparticles with guiding peptides for targeted delivery could be extended to other types of NMs for applications including genetic element delivery, targeted delivery of sensors, nutrients, or pesticides to specific plant tissues or subcellular compartments (Santana et al. 2020).

3.4 Inorganic NPs (INPs) in Plants

INPs have been studied in detail for their diversified applications during the past two decades, especially considering the metal nanoparticles (MNPs) of gold and silver; oxides of copper, zinc, iron, titanium, and graphite; core-shell nanostructures, and QDs (Wieszczycka et al. 2021).

INPs were used in water and waste water treatments for environmental protection, particularly absorbents made from magnetic iron oxide nanoparticles. Though

Fe_2O_3-NPs facilitate the separation of pollutants, their selectivity can be further improved by chemically modifying their surfaces to trace and capture specific contaminations. A variety of polymers, surfactants, ionic liquids, and organic surface modifiers are available for specific properties and applications. For instance, core-shell-NPs of iron oxide core and b-cyclodextrin (b-CD)-modified silica shell entrap the polychlorinated biphenyl from aqueous solution. The b-CD, being a cyclic derivative of starch, is a cylindrical structure with hydrophilic and hydrophobic external and internal surfaces functionalized with b-CD, offering improved adsorption properties.

NMs have also been widely used in modifying or designing new polymer membranes for water desalination, softening, and organic separation processes in which the NPs modify the membrane support or active layer. The modification of the support layer is one way to improve its mechanical and thermal properties, as well as membrane performance. NPs embedded in the active layer of the membrane affect its selectivity, separation efficiencies, and antifouling properties. For instance, the incorporation of GO into a polyamide matrix causes significant changes in hydrophilicity, surface zeta potential, thickness, and roughness of a thin-film composite membrane. GO-modified membranes also exhibit anti-biofouling properties, high water permeability, and higher salt rejection in comparison to unmodified membranes. Among the INPs used for membrane improvements, Ag-NPs combined with multi-walled carbon nanotubes (MWCNTs) have been reported as a nanohybrid membrane in which the presence of CNTs alters the physicochemical properties of the membrane support layer and improves water permeability. The functionalization of the thin-film active layer with Ag-NPs provided antibacterial and antifouling properties by influencing their wettability and fouling resistance, which depend on the membrane material. Hydrophobic polymers with high chemical and thermal stability promote membrane fouling as a result of hydrophobic interactions between pollutants and the membrane surface. In turn, the hydrophilic polymers improve membrane wettability but at the same time compromise its rejection efficiency. The use of INM's modified polymer matrix might remove the fouling and permeability challenges in various membrane separation processes. The use of modified INPs provides a better choice for detection and removal of pathogens and toxins from water and soil resources. Heavy metals contaminations, including mercury, chromium, lead, arsenic, and cadmium, can be usually removed from natural water resources via adsorption techniques. Several inorganic NMs were explored as adsorbents or adsorbent modifiers to optimize the stability and surface properties of adsorbents in terms of their hydrophobicity or hydrophilicity, surface charge, and chemical affinity to the specific pollutant. Especially, iron oxide-based NMs are known as an excellent and low-cost adsorbent of heavy metals in water and wastewater treatment applications. The Fe_2O_3-NPs, combined with their surface modifications with functional groups like carboxylic acids, amines, and organic molecules, are reported to improve selectivity and stability in water treatment, in addition to their magnetic properties, which facilitate adsorbent recovery after pollutant removal. MgO-NPs exhibit excellent heavy metal adsorption properties. A flower-like MgO-3D nanostructure demonstrated very high adsorption capacity of Pb and

Cd ions from water, which supported a new adsorption mechanism based on solid-liquid interfacial ion exchange between Mg and Pb or Cd cations. This shows that INPs offer newer pathways for excellent pathogen removal technologies. The NPs of metals like Cu, Ag, Au, and Pd; oxides of metals like Ti and Zn; and sulphides of Cd and Zn show antibacterial and/or antiviral properties caused by ROS generation and their ability to disrupt the structural integrity of the cell wall as a result of binding with peptides or other molecules on the pathogen surface. The damage caused to the microbial bodies may lead to modifications of cell membrane permeability and intracellular component leakage, as well as damage to the pathogen's DNA/RNAs (Wieszczycka et al. 2021).

NPs are widely being used in phytonanotechnology, such as mesoporous silica NPs (MSNs), carbon nanotubes (CNTs), QDs, magnetic NPs (MNPs), and metallic and metal oxide NPs. MSNs are honeycomb-like porous structures with tunable particle diameter and pore size in the nm range. Large numbers of empty channels absorb or encapsulate different bioactive molecules or agrochemicals. Plasmids containing the Green Fluorescent Protein (GFP) gene can be delivered by MSNs in the plant cells, finally triggering the expression of the target gene. MSNs conjugated with enzyme or protein can be used for genome modifications or biochemical analysis in plants.

Cylindrical CNTs with diameters between 1 and 50 nm are classified as multi-walled nanotubes (MWNTs) and single-walled nanotubes (SWNTs). QDs, nanocrystals of semiconductor materials with diameters in the range of 2–10 nm, generate fluorescence useful in subcellular imaging or labeling. MNPs comprising different magnetic materials like cobalt (Co), nickel (Ni), and iron (Fe), and their derivative compounds, are also put into a group of magnetic virus-like NPs (VNPs), carbon-coated MNPs, and other magnetic NPs that are easy to manipulate using magnetic field gradients for targeted delivery. Au and Ag-NPs are the most common NPs due to their better effectiveness in delivering biomolecules to crops. Likewise, metallic oxide NPs, consisting of ZnO, CuO, SiO_2, and TiO_2-NPs, have also been widely utilized as delivery carriers in plants' systems due to their greater light absorption, catalytic, and electrical characteristics. Numerous metal and metal oxide NPs have been applied in diverse crop management procedures, including fertilization and crop protection. Despite all the beneficial effects, attention must still be paid to design safety principles to address the concerns of the community dealing with the possible opposing influences of new NMs on the ecosystem (Jiang et al. 2021).

The unique properties of NPs make them more effective than ordinary materials like ions and molecules. In a greenhouse experiment, crops were collected after growing for 80 days in soil modified with $CuSO_4$ or CuO particles (bulk or nano). The Cu content in the roots of nano-CuO-treated crops was higher compared with bulk CuO, $CuSO_4$, or control. This indicated that CuO was more easily absorbed by crops. Time-resolved tests were performed to analyze the physiological effects and the translocation of Zn in the common bean (*Phaseolus vulgaris* L.) using X-ray absorption spectroscopy, indicating that $ZnSO_4$ and ZnO (300 nm) were more difficult to be absorbed by the roots of the bean than ZnO (40 nm). It further demonstrated that ZnO-NPs of smaller size were more easily absorbed by the crops. The highest level (78-fold) of Zn was found in ZnO-NPs-sprayed seedlings in comparison to ~ 27-fold detected in $ZnSO_4$ treated plants. It was concluded that the higher Zn accumulation with ZnO-NPs treatment was imparting better adhesion of the nano form compared to the ionic form, causing their more effective entry into the seedlings. The different transportation and accumulation features of Ag-NPs and Ag^+ were reported in rice (*Oryza sativa* L.), in which Ag-NPs were translocated and assimilated more on the root surface compared to Ag^+. The activity, aggregation, catalytic charge, crystallinity, porosity, or redox potential could possibly influence the uptake of NPs in crop seedlings. Higher NP surface activity may hasten metal component release. The electrostatic attraction, hydrophilicity, lipophilicity, and physical adsorption of NPs, influence their accumulation. For instance, neutral or positively charged NPs are more prone to agglomeration, but the NPs with negative charges are more useful for translocating in the crops (Jiang et al. 2021).

The absorption of NPs by crops can be enhanced by modifying their surface using several elements, compounds, and materials, such as iron, aminopropyl triethoxysilane, fluorescein isothiocyanate, natural organic matter, humic acid, polyvinylpyrrolidone, citrate, or polyethylene glycol. The hydrophilic protective sheet onto the NPs allows them to penetrate crops, while the protein-encapsulated NPs are more stable in crops. Nanocarriers facilitate assimilations, and nanoliposomes aid in the transport and absorption of nutrients in various crops. The chitosan molecules, for instance, decrease the agglomeration of NPs, enhance their stability, and allow them to effortlessly enter the epidermal cells of the leaves by modifying their chemical and physical properties. Moreover, the NPs' surface charges can be improved by coating, which contributes to their translocation. Recent reports found that the material of the surface coating can prevent the closure of stomata by decreasing the indiscriminate gathering of NPs, thus enhancing the absorption of NPs in crops. Moreover, the utilization of functional groups and surfactants improves the bioavailability of NPs by increasing their adhesion to the surface of leaves. For instance, hydroxyapatite can be used to alter the surface of NPs to induce their absorption in leaves by reducing their aggregation (Jiang et al. 2021).

Crop breeding is a technology to improve the genetic characteristics of crops by generating high-yield and high-quality varieties. Several conventional and molecular approaches have been used in crop breeding, including functional genomic tools, genetic selection, mutagenic breeding, physical maps, somaclonal variations, and whole-genome sequence-based approaches. Nanotechnology is a novel approach to increasing crop breeding efficiency and accuracy (Jiang et al. 2021).

Nanobiotechnology improves the efficacy of crop breeding by delivering exogenous biochemicals or nucleotides into plant cells. The barriers faced during thepenetration ofthe limited size of genetic material across the cell wall are the main barriers to the transport of external molecules into crop cells.The combination of biotechnology and nanotechnology put together offers more chances as novel means of biomolecular transport into the cells via cell walls. Recently, a nanomaterial-mediated gene delivery system was reported to have been developed. High efficiency transformationachieved in plant cells without external physical or chemical means, showed excellent applications in plant genetic engineering. For example, the foremost application of delivering NMs into crops is using MSNs to transport target genes into the leaves of tobacco (*Nicotiana benthamiana* L.). SiO_2-NPs were developed to transport DNA to crops, e.g., tobacco and maize (*Zea mays* L.), with no adverse effects. The technology of particle bombardment (the “gene gun”) has been used to transform plants since the early 1980s, using either tungsten or gold particles as DNA carriers. Using current system of NP-delivery, withreduced cell damage caused by microprojectile hits during the bombardment improves the expression efficiency of the transgenes. The DNA-coated NPs are utilized as bullets in the gene-gun technology to bombard the tissues or cells to deliver the desired genes into the target crops. SiC based plant transformation is found to transfer the sequences or fragments of DNA in various plants like maize, tobacco, rice, soybean, and cotton as an effective method. The complex of MNPs and β-glucuronidase target gene was permeated to the pollens of cotton by magnetic force with no negative effects on the viability of pollens. By pollinating with magnetofected pollen, the transgenic cotton plants were effectively selected, and exogenous genetic information was steadily inherited into offspring, achieved by selfing, successfully combined into the genome, and finally expressed. The CNT-scaffolds were applied to deliver linear or plasmid DNA to cotton, tobacco, and wheat (*Triticum aestivum* L.) leaves, causing a strong transient expression of GFP. In addition to the above-mentioned DNA delivery, NPs are also used to deliver RNA into plant cells. Chitosan NPs-embedded small interfering RNA (siRNA) delivery systems have offered a novel strategy for crop improvement by permitting the unique dominance of the target pest, as chitosan has the capability to bind with RNA and penetrate cell membranes. The double stranded RNAs (dsRNAs) carried on clay nanosheets offer defense against the cauliflower mosaic virus in tobacco leaves. The siRNA was transferred to tobacco seedlings expressing the GFP gene, resulting in a high percentage of silencing of the target gene. The CNTs-mediated platform provided RNA transfer into crop cells and protected RNA from nuclease degradation, enabling gene silencing of endogenous GFP in mutants.

The system of clustered short palindromic repeats (CRISPR) and CRISPR-associated proteins (Cas), an RNA-based guard organization in prokaryotes containing the Cas proteins and CRISPR repeat spacer arrays, has been used for genome editing in crops by traditional transformation and regeneration processes. Delivery, low HDR efficiency, species dependence, tissue culture, and regeneration are the four main challenges to be taken care of in genome editing of crops. With the characteristics of small size, high charge status, high throughput, and high tensile strength, NMs can enhance the specificity and efficiency of the CRISPR/Cas9 technique and minimize the possibility of going off-target. Contemporary developments in NMs-mediated transport of CRISPR/Cas9 single guide RNA (sgRNA) have started a novel programme of genetic engineering. Moreover, MSNs have been applied to transport Cre recombinase in the immature embryos of maize as carriers, loading the sites of the *loxP* site to recombine into the chromosomal DNA. The *loxP* fragment was integrated after the delivery of modified MSNs in crop cells. Cationic arginine gold-NPs gathered Cas9En (E-tag)-ribonucleoprotein (RNP) transport of sgRNA in cultured cell lines offered higher efficiency (~30%) of active nuclear or cytoplasmic gene editing useful in crop study. Besides current nanotechnology driven CRISPR/Cas9 technique useful in numerous crops, phytonanotechnology-based approaches are also required to take care of the other difficulties in the genome editing of crops. There are still challenges, e.g., the range of plant species that can be genetically engineered, the forms of CRISPR/Cas genome editing that can be utilized in crops, the labor and time efforts essential for crop regeneration, and the low transport efficiency (Jiang et al. 2021).

NMs, such as MSNs, Au-NPs, SiO_2-NPs, and chitosan-NPs, are known to enhance crop growth and development from the initial phases of seed germination to death or senescence in numerous crop species, including soybean, rice, wheat, peanut, tomato, potato, and onion. NMs exhibit positive effects on crops by accelerating crop breeding, promoting seed germination, increasing photosynthesis, enhancing mineral uptake, and improving crop quality and yield. The nanotechnological processes developedfor the overall growth and development of the plants are however dependent on the concentration, composition, size, and chemical and physical characteristics of NMs.

Seed germination is the first step and the most sensitive period in the life cycle of plants. Numerous studies have confirmed the use of nanotechnology to influence seed germination. Studies have shown that NMs increase water absorption and utilization by penetrating the seed coat, which can eventually improve seed germination and growth by stimulating the enzyme system. The application of Zn-NPs promotes the germination of seeds in different crop species, e.g., onion, soybean, peanut, and wheat. The use of TiO_2 and SiO_2-NPs for seed treatment, was found to substantially improve seed germination in several crops. MWCNT applications also aided seed germination in crops such as soybean, maize, peanut, wheat, tomato, garlic, and barley. The SWCNTs upregulate the expression of *SLR1* and *RTCS* genes in maize root tissues and enhances the root growth. Although many studies on the positive interactions between NMs and seed germination in crops have been published, the underlying mechanisms of NMs' superiority to traditional materials in seed germination remain unknown and require further investigation.

Photosynthesis, which transforms light energy into chemical energy, is a vital process for the growth and development of plants. Enhancing the efficiency of photosynthesis is critical for the growth of crops. It has been reported that NMs can enhance the system and enzyme activity of photosynthesis and the content of chlorophylls, which can eventually improve the overall plant growth. At the physiological level, the nano-anatase TiO_2 application enhanced the rate of photosynthesis by stimulating the Rubisco enzyme activity, which could ultimately enhance the growth and development of crops. The SiO_2-NPs increased the capacity of photosynthesis by enhancing the photosynthetic metabolism and the carbonic anhydrase enzyme activity in crops. The highest photosynthetic ability and an increase in chlorophyll content were found when the seedlings were treated with 0.5 g/L of SiO_2-NPs in wheat. A comparable effect was also reported in wheat, where TiO_2-NPs (20 nm) at 60 mg/kg increased chlorophyll content by 32.3% compared to the control. At the molecular level, TiO_2-NPs induced the *light-harvesting complex II* (*LHCII*) gene in

Arabidopsis thaliana and finally enhanced the chloroplast light absorption efficiency and thylakoid membrane LHCII content. It was reported that the Si-NPs could stimulate the *PsbY* and *HemD* gene expressions related to the biosynthesis of chlorophyll, thereby resulting in a higher photosystem II activity and an increased chlorophyll content (Jiang et al. 2021).

NPs could influence the yield of crops by changing the physiology and biochemistry of plants. Fe_3O_4-NPs maintained iron homeostasis, decreased lipid peroxidation, and induced ferritin content in maize. Crops grown in soil modified (hydrophilic or hydrophobic coating) or unmodified with nano-TiO_2 for 2 months displayed that inorganic carbohydrate production, mineral uptake (P, Se, Mn, Fe, Mg, Ca, and Cu), chlorophyll, enzyme activity, and plant growth were significantly enhanced with the application of coated NPs. Similarly, foliar applications of Mn, Fe_2O_3, and Mo-NPs also increased the crop yield. Moreover, various NPs have been used to enhance the dietary value and nutritional components of fruits or food grains. Foliar application of Cu-NPs increased the antioxidant enzyme activity, fruit firmness, and vitamin content, thus enhancing the freshness and quality of tomatoes. At the molecular level, The Ag-NPs are found affecting the expression of *VIN3*, *VRN2*, *FRI*, and *FLC* in the vernalization pathway and cause the downregulation of key flowering control genes *AP1* and *FT*, thereby delaying flowering. The application of different NPs influences the quality and yield of crop seeds or fruits, which completely depend on the mode of treatment, properties, size, and type of NPs used. Further research into dose dependence, long-term exposure effects, and molecular studies such as metabolomics or proteomics could lead to a better understanding of how NPs affect the quality and yield of seeds or fruits.

The importance of phytonanotechnology is reflected in the crop-derived foods in terms of their roles in improved food quality, e.g., taste and nutritional value. The application of nanotechnology offers numerous ways to improve food taste. As compared to the larger materials, which usually degrade over a longer duration, the smaller NPs deliver favorable ways of enhancing food bioavailability because their nearly subcellular size contributes to a higher level of bioavailability. Several metal oxides like SiO_2 and TiO_2-NPs have been utilized as flow agents or colorants in food products. SiO_2-NPs were used to deliver flavors or fragrances into the food items. The novel technology of nano-encapsulation has been widely utilized to provide a balance between culinary aspects by increasing flavor preservation and release. The encapsulation of ferritin nanocages could increase the thermal stability and solubility of ferritin.

Large numbers of bioactive substances like vitamins, proteins, carbohydrates, and lipids are more sensitive to the enzyme activity of the duodenum and stomach and the high acidic environment. Nano-encapsulation permits the bioactive substances to assimilate readily in food products and allows them to struggle with such opposing conditions, which is difficult to attain in the non-capsulated forms because of the low water solubility of the bioactive substances. The diverse tools that have been used to encapsulate compounds in tiny shapes to transport nutrients more efficiently like antioxidants or proteins for health or nutritional assistance include nano-emulsification, nanostructures, and nanocomposites. Polymeric NPs are reported to be appropriate for bioactive substance encapsulation (such as vitamins or flavonoids) to transport and protect the bioactive substances for different purposes. NPs-based tiny edible capsules are being developed to improve the delivery of fragile micronutrients, vitamins, and medicines in a daily-use food item, offering a significant benefit to human health.

Food safety has become a worldwide issue because of increasing food demand and compromised crop yields resulting from climate change, soil degradation, and crop disease proliferation. Nanotechnological processes like nanofabrication and nanoencapsulation, are expected to provide new value-added solutions for the fortification of foods with bioactives and targeted controlled release in the gut to compensate for the food demand-supply objective.

A required packaging material must have moisture and gas permeability along with biodegradability and strength. Microcapsules of vegetable oils have been used in postharvest preservation for a long time, and nano-facilitated "active" and "smart" food packaging are anticipated to provide numerous advantages over conventional techniques with improved antimicrobial films, barrier properties, and mechanical strength. Nanomaterials could also support food preservation aspects by being added directly into a food matrix or food contact materials such as packaging. The increasing application of nanotechnology offers numerous selections to extend the shelf life of food for longer use. The area of food preservation is expected to benefit significantlyby nanotechnologyespeciallynotedinreducedspoilage.Nanoliposomes, nanoemulsions, edible coatings, and other different kinds of NPs have been widely used in the postharvest preservation of crop-derived foods.

Nanoliposomes have been certified to be valid in the targeted transport of antimicrobial compounds, nutraceuticals, nutrients, and vitamins owing to their small size with a larger area for surface contact. It has been found that many factors are influencing the fusion of liposomes with bacterial cells, like the pH and temperature of the bacterium's surface, the presence of divalent cations, and the characteristics of the bacterial membrane. The delivery of negatively charged liposomes into the cells is promoted by nonspecific receptors, followed by particle recognition, adhesion, and ultimately, endocytosis. The lipid composition of the bacterial membrane plays an important role in the fusion of liposomes with the bacterium.

Nanoemulsions are mixtures of different immiscible liquids that are stabilized by a surfactant (e.g., modified starches, proteins, and lipids) and have an average droplet size of 20 to 200 nm. The food-grade surfactants including lecithins, sugar esters, and polysorbates emulsifying agents are useful in the postharvest preservation of crops with better stabilization of the nanoemulsions through adequate responses to environmental stresses, such as enhanced loading capacity, steric hindrance, and repulsive electrostatic interactions. The characteristics of nanoemulsions make them an appropriate intermediator for the transport of hydrophobic drugs, bioactive molecules, and nutraceuticals that encapsulate hydrophobic antioxidant ingredients (Jiang et al. 2021).

Edible coatings have been reported to extend the shelf life of perishable crop-derived foods, particularly vegetables or fruits. Edible coatings are traditionally prepared using thin materials applied with food-grade substances that are utilized on the surface of vegetables and fruits and become a part of the crop-derived foods that stay on the outer surface during consumption. The edible coatings contribute to postharvest preservation by controlling carbon dioxide and oxygen permeability, decreasing moisture loss, and thus decreasing the rates of solute respiration, oxidation, and migration without compromising the quantity or quality of the crop-derived foods. Edible coatings have good performance to play as a carrier vehicle for antimicrobial agents, micronutrients, flavoring compounds, colored pigments, and anti-browning mediums that assist in increasing the validity period of crop-derived foods by preventing the survival of the pathogens on the outer surface of fresh vegetables and fruits during the food spoilage.

Nanocomposites have widely been usedin food packaging and preservation because of their antifungal or antibacterial features. Ag-NPs' antimicrobial property has been widely used for postharvest preservation due to their inactivation of food spoilage microorganisms. Special properties of Ag, such as low volatility, stability at high temperatures, and toxicity against various types of microbes, make it a beneficial choice for application in postharvest preservation. Nanocomposites applied with Ag-NPs have been utilized to prolong the shelf life of different kinds of crop-derived foods. The mechanism of Ag-NPs activity is mediated via the release of Ag^+ that adhere to the cell surface of pathogens, disrupts the cell membrane of pathogens, damagestheir DNAs, and causes cell death. The primary mechanism of Ag-NPs has been extracted in three routes: interruption of ATP production and DNA replication by the Ag^+ uptake, production of ROS by Ag-NPs and Ag^+ in the cells, and finally deterioration of the membrane of the cells by Ag-NPs. Because of their excellent antimicrobial characteristics against microorganisms, TiO_2-NPs are extensively utilized in postharvest preservation. Photocatalysis of TiO_2 inactivates food spoilage bacteria by inducing the lipid peroxidation of phospholipids in microorganisms' cell membranes. The antimicrobial mechanism of TiO_2-NPs can be described in three routes: ROS production upon the activation by the UV and visible wavelengths of light, oxidative stress and lipid peroxidation in cells, and the death of cells. Despite this, more research is needed to investigate the underlying mechanisms of interactions between NPs and crops to increase shelf-life and delay ripening.

3.5 Chapter Summary

Despite great improvements introduced by nanotechnology in plant genetics and crop breeding, the delivery of exogenous enzymes or DNA for genome editing is still not that easy. Based on evidence found in plant cells, soft materials, like polymeric nanostructures and nanogels, can be utilized as promising substances to advance novel approaches for genome editing and the controlled release of biomolecules in crops. The problems of phytonanotechnology can be taken care of by initiating multidisciplinary studies for the synthesizing and designing of intelligent NMs. Strong joint collaborative efforts put in by competent researchers, including chemists, biochemists, engineers, geneticists, and botanists, would certainly give rise to novel developments in phytonanotechnology. For improved crop growth and related developments, the current applications propose that more studies are needed to improve the sustainability of agricultural systems. Future studies involving open-field trials may further benefit from recognizing the mechanism of NPs' action on crops. The NPs discussed above need to be designed to meet the requirements of their application methods (soil or foliar) so that they can have a high impact and ensure better crop quality. In the meantime, excessive use of these NMs might pollute the environment, and therefore special care must be taken while working with NMs in plant systems. However, it is an undeniable fact that the positive functions of NPs have shown great potential in various aspects of agricultural systems, starting from germination to postharvest.

Further Reading

Jiang M, Song Y, Kanwar MK et al (2021) Phytonanotechnology applications in modern agriculture. J Nanobiotechnol. https://doi.org/10.1186/s12951-021-01176-w

Santana I, Wu H, Hu P et al (2020) Targeted delivery of nanomaterials with chemical cargoes in plants enabled by a biorecognition motif. Nat Commun. https://doi.org/10.1038/s41467-020-15731-w

Sanzari I, Leone A, Ambrosone A (2019) Nanotechnology in plant science: to make a long story short. Front Bioeng Biotechnol. https://doi.org/10.3389/fbioe.2019.00120

Wieszczycka K, Staszak K, Woźniak-Budych MJ et al (2021) Surface functionalization – the way for advanced applications of smart materials. Coord Chem Rev. https://doi.org/10.1016/j.ccr.2021.213846

Chapter 4
CNMs and CDs in Plant Growth

4.1 Introduction

Carbon-based nanomaterials (CNMs) comprising carbon black, fullerenes, carbon nanohorns, carbon nanodiamonds, carbon nanofibers, carbon nanotubes, and carbon dots possessing novel properties have been explored in a variety of sensing applications such as disease diagnosis, metal ion detection, food safety, and environmental regulations. Some of these carbon nanomaterials are increasingly being used as fluorescent probes for bioimaging purposes. Moreover, they have also been employed as nanocarriers for drug or gene delivery, besides acting as therapeutic agents in photodynamic and photothermal therapies (Mansuriya and Altintas 2021).

4.2 CNMs and Plants Interaction

A number of different CNMs, namely fullerenes, carbon-NPs, fullerol, and SWCNT/MWCNTs, are currently gaining importance in agriculture due to their possible applications in regulating plant growth and antimicrobial properties. The influences of CNMs have shown positive as well as negative effects (Mukherjee et al. 2016).

4.2.1 Plant Growth

Earlier studies, while evaluating the CNM accumulation in plants, exhibited a significant increase (~17%) in ryegrass root length upon exposure to 2000 mg/L MWCNT as compared to untreated controls. The toxicities of uncoated and poly-3-aminobenzenesulfonic acid-coated SWCNTs were compared in six crop species:

A. Nazeer et al., *Targeted Delivery of Nanopesticides and Nanofertilizers in Sustainable Agricultural Farming*, Nanotechnology in the Life Sciences,
https://doi.org/10.1007/978-3-031-41333-9_4

cucumber, carrot, onion, tomato, cabbage, lettuce, and uncoated CNTs. The results showed increased root length in onion and cucumber as compared to the coated CNTs. An inverse relationship was noted between the exposure time and the extent of root elongation, with no evidence of root internalization under microscopic examinations, and hence only surface adsorption took place. Upon exposure to wheat, CNTs were not only adsorbed onto the root surface but also penetrated the root epidermal cells and accumulated within the tissues. In contrast, the exposure of citrate-coated, water-soluble CNTs in gram showed CNT-internalization in electron microscopic examinations. There have been several other reports of CNT exposure enhancing plant growth. The effect of pristine (~30 nm) and oxidized MWCNT (~20 nm) on mustard exposures reported enhanced germination as well as increased root and shoot growth. However, the germination started decreasing at higher exposure levels. SWCNT and MWCNT exposures were noted to enhance the total fresh biomass of tomato seeds by 75 and 110%, respectively, as compared to activated carbon and graphene. In a follow-up study, the effects of MWCNT and activated carbon exposure on tobacco cells demonstrated 55–64% higher levels at 5–500 μg/mL MWCNT compared to untreated controls. Activated carbon enhanced cell growth (16%) at low concentrations (5 μg/mL), but growth was suppressed by ~25% at higher exposures (100–500 μg/mL). These observations were attributed to the upregulation of several genes upon CNT exposure, including aquaporin (NtPIP1), CycB, and NtLRX, which are involved in water transport, cell wall formation, and cell division. Similarly, there are reports regarding ~50 and 32% increase in root length of wheat seedlings after 3 and 7 days of exposure to MWCNT, respectively. The effect of MWCNT exposures (10–11 days @ 50, 100, and 200 μg/mL) on the germination and growth of soybean, corn, and barley in an agar medium exhibited 50% and 90% increases in barley and soybean, respectively, and in corn germination rates, respectively, compared to untreated controls. Corn exposed to 5–60 mg/L MWCNTs for 7 days in an agar medium showed increased plant-fresh biomass and nutrient uptake compared to the controls. Many studies have reported positive impacts of CNT exposure, although most of these observations are from short-term studies with high levels of CNT exposure, mostly in artificial growth media. The relevance of these findings to actual agricultural conditions remains unknown, but the consistency of the findings across plant species upon CNT treatment clearly warrants further investigation (Mukherjee et al. 2016).

Enhanced root growth (10×) of wheat upon 10 days of exposure to 150 mg/L water-soluble CNDs as compared to controls. Similarly, growth enhancement in gram plants treated with water-soluble carbon nano-onions exhibited exposure-dependent growth when compared to untreated controls. Water-soluble carbon nanoparticles such as fullerols [$C_{60}(OH)_{20}$] are another candidate for carbon-based plant growth enhancement in the biomass and phytomedicinal content of bitter melon. Plants treated with different concentrations of fullerol exhibited a 54 and 128% increase in biomass and fruit yield, respectively.

In the fruit tissues, fullerol exposure significantly increased the content of cucurbitacin-B (74%), lycopene (82%), and other anticancer and anti-diabetic molecules by 20% and 91%, respectively. The concentration-dependent effects of water-soluble CNPs on wheat showed optimum growth at 50 mg/L treatment, with root and shoot lengths increasing up to three times compared to untreated controls. Carbon nanohorns have also been reported to have a positive impact on the growth of terrestrial plants. CNHs, spherical structures with disordered single-layered graphene sheets, when exposed to tobacco cells, showed a 78% increase in growth of cultured tobacco cells at 100 μg/mL while no significant effects were noted at 25 μg/mL, as compared to controls. In a recent study of treated tomato seeds in cotton-cushioned glass bottles with 40 μg/mL graphene, increased germination rates were observed compared to untreated controls.

Besides, the fate and transport of several organic co-contaminants were reported, showing fullerene-dependent uptake of trichloroethylene by eastern cottonwood increasing by 26 and 82% at 2 and 15 mg/L fullerene, respectively. Conversely, the effects of C60 fullerene on the bioaccumulation of p,p'-DDE by zucchini, soybean, and tomato in vermiculite exhibited differential uptake across different plant species. In zucchini and soybean, a 29% increase and a 48% decrease in p,p'-DDE uptake were reported upon fullerene exposure, respectively, but no effect was evidenced in tomato. Reduced accumulations (21–80%) of chlordane and DDx (DDT + metabolites) has been reported in four crops (zucchini, corn, tomato, and soybean) upon MWCNT co-exposure. However, C60 co-exposure showed mixed effects, ranging from increased chlordane uptake of 34.9% (soybean/tomato) to complete loss of DDX uptake (tomato/corn). Species dependent uptake of DDE was also noted in lettuce, influencing the uptake of chlordane components and p,p0-DDE. These findings indicate that CNMs have the potential to improve plant growth, nutrient uptake, seed germination, and fruit quality.

CNTs are the most extensively studied and have shown promising positive effects, with low to moderately high doses of CNTs improving overall plant growth. CNTs and fullerenes were also found to have the secondary positive effect of reducing the accumulation of pesticides in selected plant species. However, across all studies, the beneficial responses are largely dependent on plant species, nature of the growth medium, CNM type and concentration, and growth conditions.

Publications in related areas reveal a number of adverse effects on the plants from exposure to a range of CNMs. For instance, the toxicity was found to be largely dependent on CNM concentrations, growth and exposure conditions, and plant species. However, with soil-based studies in fields being scarce, the extrapolated findings might not describe the field conditions (Mukherjee et al. 2016).

The effect of MWCNT exposure under hydroponic conditions on zucchini reported a biomass reduction when compared to control and bulk carbon. A separate hydroponic study showed dose-dependent toxicity of MWCNTs in mustard, where oxidized MWCNT exerted more negative effects than pristine MWCNTs. At a "high" exposure concentration, both pristine and oxidized MWCNT caused toxicity, reducing germination by 4.4% and 7.6% and dry biomass by 1.6 and 2.2-fold, respectively, as compared to the lowest concentration. At its highest concentration (2000 mg/L), graphene significantly reduced plant growth (up to 78%), biomass (up to 88%), the number and size of leaves (up to 53% and 91%, respectively), and increased ROS production and necrotic symptoms in all plants except lettuce.

Significant increase in fragmented nuclei, membrane damage, ROS generation, mitochondrial dysfunction, and induced cell death were observed upon exposure, as noted in the case of graphene oxide (GO) on the fava bean glutathione redox system, a major determinant of cellular redox homeostasis that showed concentration-specific stress responses as well as decreased oxidative enzyme activity. Subsequently, no toxicity was found below 800 mg/L of GO; the highest concentration (1600 mg/L) resulted in growth reduction, decreased anti-oxidative enzyme activity, and greater electrolyte leakage (Mukherjee et al. 2016).

The fullerene toxicity to higher terrestrial plants was studied to explain the effects of water-soluble carboxyfullerenes in tobacco BY-2 cells exposed to 0.01 mg/mL for 3 days in cell culture medium, showing cell boundary disruption and growth inhibition. In a co-exposure with GO and arsenate, the adverse effects of As(V) in wheat seeds got enhanced resulting in reduced fresh mass, shoot length, and chlorophyll content. The activities of peroxidase (POD) and superoxide dismutase (SOD) were increased in a concentration-dependent manner.

The negative effects of CNMs are specific to growth conditions and plant species. More research needs to be done to uncover the mechanisms of CNM-plant interactions, and these fundamental studies should be conducted under a range of growth conditions and with a large number of plant species.

4.2.2 Interactions with Soil Microbes

The protection of soil microbial biomass and the existing diversity is a major challenge in agriculture. Currently, limited information is available on the interaction between CNTs and the soil microbial community. CNMs might be directly toxic to soil microorganisms and alter the bioavailability of nutrients or increase or reduce the toxicity of organic compounds and/or toxins.

Although limited information is available, CNMs seem to be generally toxic to soil microbes. Nonetheless, a few studies on soil microorganisms have revealed neutral or positive biological effects. For example, the impact of fullerenes (C60) on soil microbial community populations was evaluated using total phospholipid-derived phosphate after soil treatment with 1 and 1000 μg C60/g soil for 180 days; the results showed that fullerenes had no impact on the structure or function of the soil microbial community or on soil enzymatic activities. Similarly, after several months of exposure to fullerene (C60) concentrations up to 50,000 mg/kg, anaerobic wastewater treatment sludge showed no significant effects on microbial community activity. The diverse effects on sandy loam soil microbial communities after 90 days of exposure to MWCNT exhibited no observable effects on soil microbial composition and enzymatic activities. However, at 10,000 mg/kg a mixed response was observed in terms of reduced abundance of selected bacterial species like *Waddlia*, *Holophaga*, *Derxia*, and *Opitutus*. The amount of polycyclic aromatic hydrocarbon (PAH)-degrading organisms like *Cellulomonas*, *Rhodococcus*, *Pseudomonas*, and *Nocardioides* was found to have increased. These findings suggest that as soil-MWCNT concentration increases, a potential shift toward more stress-tolerant organisms occurs, though the findings are too limited to be conclusive (Mukherjee et al. 2016). GO exposure (0.1 g/L) enhanced the activity of anaerobic ammonium-oxidizing bacteria by 10%. A dose-specific enhanced production of protein and carbohydrate was reported with GO concentrations of 0.05–0.1 mg/mL.

The mechanisms responsible for these interactions are unknown; however, this limited literature does suggest that under certain exposures, CNMs might have neutral or perhaps modestly beneficial effects on microbial communities.

Reports exist to confirm that CNTs interact strongly with bacterial cell membranes via strong electrostatic interactions, causing membrane ruptures by oxidative stress and/or physical damage. For example, studies with gram-negative *Escherichia coli*, *Pseudomonas aeruginosa*, and gram-positive *Staphylococcus aureus* and *Bacillus subtilis* strains were used in evaluating the toxicity of SWCNT dispersed (5 g/mL) in Tween 20 saline solution compared to SWCNT aggregates in saline solution. Dispersed SWCNT demonstrated greater antibacterial activity against gram-positive bacteria. The SWCNTs dispersed in Tween 20 saline solution showed antibacterial activity of 58.1 ± 5.0% for *E. coli*, 65.1 ± 1.8% for *P. aeruginosa*, 87.5 ± 6.5% for *B. subtilis*, and 85.6 ± 5.3% for *S. aureus*. On the other hand, SWCNT aggregates in saline solution showed reduced antibacterial activity of 33.8 ± 4.0% for *E. coli*, 27.7 ± 5.9% for *P. aeruginosa*, 53.9 ± 2.8% for *B. subtilis*, and 50.3 ± 3.5% for *S. aureus* (Mukherjee et al. 2016).

The antimicrobial activity of SWCNT was reported on bacteria deposited on surfaces and in suspension, showing the effects of surface coating with SWCNTs on the soil microbial community under low and high organic matter concentrations. Upon 6000 μg/g of PEGylated SWNTs exposed for 6 weeks showed some microbial community shift keeping the total biomass unaffected. Studies with purified SWNTs at concentrations varying from 1 to 50 μg/mL exhibited strong antimicrobial properties against *E. coli,* causing loss of viability and subsequent inactivation. Another important factor that influenced the toxicity of CNTs to bacteria was the presence of residual impurities. Commercial CNTs are synthesized with strong acids and contain up to 4.5–15% of metals such as cobalt (Co), iron (Fe), nickel (Ni), and yttrium (Y) and other impurities, which may exert toxic effects on microbes. Another possible mechanism of CNT toxicity is ROS induction, which might then directly interact with organelles to induce DNA damage or protein inactivation that results in apoptosis and cell death. Other studies have found that carbon-based materials have a wide range of toxic effects on soil bacteria. For example, protozoans and bacterial total respiration, biomass, and diversity showed a threefold decrease in the number of fast-growing bacteria after C60 exposure at 0–50 mg/kg dry soil. The protozoan population was only slightly affected at the highest fullerene concentration.

The aqueous solution of 0.5 mg/L of C60 showed an alteration in *P. putida*'s (Gram negative) phase transition temperatures and levels of unsaturated fatty acids in the bacterial membrane. The effect of carboxyl-functionalized SWCNTs on soil bacterial and fungal communities was studied with amendments of 0, 250, and 500 mg/kg functionalized SWCNT for 14 days, and the populations were monitored over time. After 3 days, the number of colony-forming units was significantly decreased, but the population had recovered after 14 days. Alternatively, higher doses of SWNTs had a similar biomass loss at 3 days, but the fungal community was unable to recover even after 14 days. A negative correlation with SWCNT concentration in soil was reported at relatively low concentrations (0.03–1 mg/g) in terms of the biomass of gram-positive and gram-negative bacteria, as well as fungal populations.

The enzyme activity decreased at 30 min of incubation, and for most enzymes, the suppression persisted until day 11. In both soils, microbial biomass and extracellular enzyme activity decreased (up to ~50%) and reductions were more notable at higher exposure concentrations (500 and 5000 μg/g). Recently, the soil bacterial community composition was found to be influenced by the functionalized MWCNT at the highest concentrations; however, after 8 weeks, there was no effect on the bacterial diversity with either type of nanotube. The impact of GO at 0.5–1 mg/kg noted a reduction of up to 50% in the activity of select key soil enzymes, including xylosidase, 1,4-ß-N-acetylglucosaminidase and phosphatase, after 21 days of exposure.

These observations suggest that CNMs may have a significant negative effect on soil microbial communities. Less information is available on the broader impacts of these adverse effects, including the implications for symbiotic or co-habitating terrestrial plant species.

Considering the varied applications of CNTs and their scope in agriculture, it is necessary to design realistic exposure conditions to investigate CNT fate and effects. This includes testing environmentally realistic concentrations under relevant conditions. But so far, neither of these parameters has been achieved. Since laboratory based recent studies are likely to lead to inaccurate predictions of CNT fate and effects, assessment under "realistic" conditions and in more complex and relevant biological conditions is needed to appropriately assess the risk from the widespread uses of CNTs in agricultural settings.

Not surprisingly, the majority of research on CNM-toxicity has been related to the majority of existing publications, which focuses on the straightforward evaluation of physiological and biochemical effects. The interaction among various environmental parameters like growth media, exposure time, receptor or species differences, and CNM type, synthesis conditions, and concentrations is very complex, and efforts to ascertain consistent or broad trends with regard to the material fate and effects are rather difficult to assess. The artefacts from sample preparation, storage, and/or experimental design influence the results, leading to biased or inaccurate data interpretation.

Little progress has been made on understanding the underlying mechanisms of toxicity at the molecular/genetic level due to, e.g., current limitations to detect CNMs in complex growth media; the unavailability of *in situ* real-time monitoring techniques to track them inside the living tissues and different environmental components; the complex, heterogeneous, and inconsistent nature of the soil matrix; the lack of understanding their distributions in the environment; and the species and growth condition dependence of the induced toxicity. Additionally, in the growth medium during exposure, the NM surface properties may get altered, which subsequently might cause changes in material bioavailability and toxicological outcomes (Mukherjee et al. 2016).

On the other side, the impact of agricultural feedstock on nanomaterial synthesis, structure, and reactivity still remains elusive. There are few reports where organic extracts from agricultural plants like grass and tea, among others, has been successfully used as the alternative carbon sources for CNM green-synthesis *in-situ*. Nonetheless, the mechanistic understanding of this approach remains mostly unknown and needs further investigation involvingthree major modes of action for NMs interaction via dissolution, direct contact, and co-transport of other contaminants.

Pristine CNMs resist dissolution; however, the functionalized forms such as carboxylated CNTs and C60-OH are more likely to dissolve in the growth media. Toxic heavy metals like Ni, Co, and Fe associated with CNMs could also alter the overall toxicological profile. Therefore, a comprehensive knowledge of the dissolution or degradation kinetics of the CNMs in the environment is necessary to understand the underlying mechanistic or biochemical pathways. Additionally, direct physical contact has been shown to change plant physiological and biochemical parameters such as biomass production, growth, germination rates, and ROS production.

Several studies have shown CNM-mediated alterations of pesticide uptake/toxicity in terrestrial plants, including zucchini, soybean, tomato, and lettuce. Unlike metallic NMs, there are very few analytical methods available for detecting CNMs in the environment. The relevant concentration of CNMs can go down to as low as 0.05 μg/kg. The available analytical techniques, including TEM, SEM, AFM, and Raman spectroscopy, have been helpful in determining the presence of CNMs inside the plant tissue. These techniques are capable of identifying compounds up to the sub-nanometre level. However, imaging techniques like TEM and SEM need extensive sample preparation, including solid or liquid extraction, fixation, and staining, which can alter the morphology of individual CNMs or their clusters. Softer imaging techniques like Raman spectroscopy might lead to surface modification of the CNMs. This shows the technological limitations and makes it imperative to develop advanced imaging technologies with very little or no prior sample preparation without compromising the sub-nm resolution.

Another emerging category of CNMs is called carbon dots (CDs), which are photoluminescent species of <10 nm size, although there are reports that the CDs could be as large as 60 nm in size. The structure of CDs is comprised of sp^2 and sp^3 bonded C-atoms with a large number of polymer chains or functional groups attached to their surfaces. CDs exhibit excellent electron conductivity, photobleaching and photo blinking properties, high photoluminescent quantum yield, fluorescence property, resistance to photo-decomposition, alterable excitation and emission attributes, increased electro-catalytic activity, good solubility in aqueous media, excellent biocompatibility, long-term chemical stability, cost-effectiveness, and negligible toxicity. CDs are functionalized easily due to the abundant functional groups present on their surface, and they exhibit remarkable sensing features such as specific, selective, and multiplex detectability. The presence of abundant functional groups like amine, carboxyl, hydroxyl, and others or polymer chains on the surface of CDs results in excellent solubility in aqueous solutions, as well as making them convenient for being functionalized with other nanomaterials. The carbon skeleton of CDs displays either an amorphous carbon form or graphite lattice due to the variation in their degree of carbonization. In addition, the physicochemical characteristics of CDs are tunable based on their intended usage or application. In the area of biomedicine and sensing technology, CDs are usually selected as the transducing elements or as electrode modifiers, either in combination with other nanomaterials or as an individual nanomaterial for the desired sensing applications. CDs possess a combination of electrical, mechanical, and optical characteristics to introduce miniaturized sensors with outstanding attributes for point-of-care testing (POCT). CDs, as potent imaging agents, play a significant role in the real-time imaging of either certain cells, tissues, organs, or combinations thereof that help in the accurate diagnosis of various forms of cancer and several diseases. Moreover, they have the tendency to efficiently deliver genes or drugs by serving as nanocarriers, as well as being highly capable theranostic agents for various phototherapies like photodynamic and photothermal therapies (Mansuriya and Altintas 2021).

CDs are generally classified as per their core structure, surface functional groups, and properties of the carbon quantum dots (CQDs), graphene quantum dots (GQDs), carbon nanodots (CNDs), and carbonized polymer dots (CPDs). CQDs are primarily nanospheres that are crystalline in nature and have a large number of chemical groups attached to them that impart intrinsic state luminescence and the quantum confinement effect. Ultra-small fragments of graphene (GQDs) are anisotropic in nature and constitute mono- or multiple layers of graphene sheets with graphene networks in their configuration. GQDs have quantum confinement and edge effects because of the existence of various chemical functionalities on their edges or within their interlayer defects. The CNDs are highly carbonized with edge effects, but without disclosing the crystalline or polymeric structures. Besides, CNDs lack the ability to display the quantum confinement effect. The CPDs are nanohybrids of carbon and aggregated polymers with a central carbonized-core enveloped by either polymeric chains or functional groups.

The outstanding properties of CDs are extensively used in several biomedical areas, including biosensing, bioimaging, and therapy development. The electrochemical and optical properties of CDs are briefly described for applications, particularly in biosensing, bioimaging, drug/gene delivery, and photodynamic and photothermal therapy. CDs are currently being used in the area of electrochemistry and electrocatalysis because of the following advantages:

- CDs exhibit exceptional charge transferability, enhanced electroconductivity, a larger effective surface area, and lesser toxicity, besides being cost-effective.
- Abundant functional groups like hydroxyl, carboxyl, and amine, present onto-their surfacesprovide a large number of sites for the surface modification, as well as for the enhanced electrocatalytic activity by accelerating the intermolecular electroconductivity,

- Heteroatom-doped CDs involving N, P, S, and B exhibit significantly improved electronic attributes due to the intramolecular charge transferability.
- CDs enhance the electrocatalytic processes during the electrochemical reactions such as the oxygen evolution reaction (OER), hydrogen evolution reaction (HER), oxygen reduction reaction (ORR), and alcohol oxidation reaction (AOR).

The merits of CDs described make them ideal electrocatalytic agents to serve desired electrochemical applications.

CDs display superior electrical conductivity, and when they are employed as electrocatalysts during the catalytic reactions, the Schottky barrier formed at an electrolyte-catalyst junction can be readily removed, which can thereby confirm the effective energy transformation. Moreover, owing to their excellent electrical conductivity, CDs could rapidly transfer electrons during electrochemical reactions.

Doping of CDs with nitrogen, phosphorous, sulphur, and boron exhibits changes in their chemical structure, whereas the electric charge can be efficiently transferred from the adjacent carbon atoms. The heteroatom-doped CDs show exceptional electrochemical performance due to the enhancement of intrinsic activity of surface functional sites, the distortion of their electronic configuration, the tuning of local densities, as well as the acceleration of adsorption and desorption phenomena.

Due to the presence of large numbers of active functional groups on the surfaces as well as long-term chemical stability in a wide range of solvents, they are proven to be the ideal CNMs in terms of the improved chemical stability of hybrid catalysts. CDs are used as supporting materials during the preparation of their hybrid nanocomposites with metals and metal oxides, which could prevent agglomeration and thereby increase the electrocatalytic activity. CDs demonstrate good stability in aqueous media due to the electrostatic stabilization, which can facilitate the stability of hybrid catalysts. Because CDs interact strongly with catalysts via electrostatic interaction, they can significantly improve stability.

Excellent electroconductivity, numerous defect sites, and active edges, as well as the large surface area-to-volume ratios of CDs make them as highly active center of interactions. CDs being fused together with conductive materials can dramatically facilitate their electrochemical performance and characteristics.

Fluorescent CDs have been deployed in diverse health care applications, especially in the fields of biosensing, bioimaging, and therapy development. It is critical to investigate and comprehend the optical properties of CDs in order to prepare a variety of CDs for various bio-applications.

Due to the $\pi - \pi^*$ transition of C=C bonds in CDs, they exhibit optical absorption in the short-wavelength region. CDs exhibit intense optical absorbance from 260 to 320 nm in the UV region. The absorption range may vary as a function of the type of CD, due to the surface functional groups as well as their surface passivation.

4.2.3 Fluorescent CDs

The up- and down-conversion fluorescence of CDs is defined in terms of the difference between the excitation and emission wavelengths. The up-conversion fluorescence property can be observed in the CDs that are synthesized through ultrasonic treatment. A larger excitation wavelength results in the reduction of background autofluorescence, which is significant for the bioimaging application.

The luminescent properties of CDs are yet to be deeply investigated. The fluorescence of CDs is the result interactions among multi-emissive centres, free zigzag sites, self-trapped excitons, quantum confinement effects, special edge defects, their conjugated structures, and surface states. The fluorescence of CDs is assigned to the presence of an electron-hole pair of excitons. As the size of CDs increases, their energy gap decreases. Therefore, the fluorescence property of CDs can be regulated by altering their quantum confinement effect. The surface state phenomenon due to the existence of surface functional groups and surface oxidation is another mechanism for the origin of CDs' fluorescence. The surface oxidation caused by oxygen-containing groups at the edges of CDs is responsible for creating the surface defects that result in the fluorescence.

Different fluorescence emission properties of CDs are noted to depend on excitation wavelength and it is achieved by regulating several physicochemical parameters during CDs' synthesis. CD fluorescence, for example, is heavily influenced by pH, concentration, and temperature. The pH-dependent emission is because of the functional group protonation and deprotonation on their surfaces; the concentration-dependent fluorescence is due to the surface state emission; whereas the temperature-dependent emission is the result of nonradiative decay occurring at the surface of CDs.

Process of fluorescence bioimaging or biosensing requires long emission lifetimes and a stable fluorescence signal. This can be achieved with the help of CDs, since they have the tendency to produce stable signals when stored in an aqueous environment. CDs can emit strong fluorescence for about a year. Generally, CDs are resistant to a broad pH range (i.e., from 3 to 12); therefore, they are immune to photobleaching.

The room-temperature phosphorescence (RTP) property of CDs is of great importance due to its long lifetime. Out of two factors to be considered for RTP, the first one involves the suppression of non-radiative transitions by restricting rotation and vibration, while the second one facilitates the inter-system crossing ability by enriching the spin-orbit coupling involving the transition metals.

RTP is possible to achieve by employing CDs with enormously cross-linked structures containing non-conjugated groups. The production of RTP in aqueous media is relatively challenging since phosphorescence quenching is commonly observed in water because of solvent-assisted relaxation as well as the presence of dissolved oxygen.

In chemiluminescence (CL), the light is produced via a chemical reaction under appropriate conditions in a redox reaction generating CL in aqueous solvents, where the unstable products are produced from intermediate radicals. CL from CDs arises either due to the excitation after direct oxidation or through the enhancement or inhibition of their luminescence.

CDs are able to emit photons in the visible region under electrical excitation, owing to the enhanced electron transfer due to a large amount of sp^2 carbon in CDs, causing a stable ECL.

4.2.4 Strategies for CDs Synthesis

During the synthesis of CDs, several functional groups, such as amine, epoxy, ether, carbonyl, carboxyl, and hydroxyl, can be incorporated. Besides, the surface of CDs can be easily functionalized by doping them with heteroatoms like N, P, S, and B using various biological, polymeric, and organic materials. The characteristics of CDs can be controlled by varying the size and extent of the surface functional groups, either by adapting different synthesis techniques or using other precursors. The modification of CDs is crucial for achieving their considerable surface attributes for solvency as well as their favourable applications. The quantum yield (QY) of CDs is possible to improve either making changes while performing their syntheses or even after their preparation. Significant investigations have been carried out recently to prepare CDs with high QY for better bioapplications. Nonetheless, the high QY of CDs and their biocompatibility compete with each other, as well as it is quite difficult to regulate both of them individually. For the achievement of higher biocompatibility, the surface passivation of CDs can be improved, which may result in a reduction in their photoluminescence intensity or vice versa. Despite these issues, it is still unclear how CDs function as excellent fluorophores for serving various health care applications.

Innumerable food-containing carbon sources can be employed to synthesize CDs with different QYs, such as yoghurt, honey, banana, pomegranate, leaves, sugar beet molasses, eggs, rice bran, garlic, coffee beans, soybeans, coconut shell, tea leaves, and grass, to name a few. Additionally, various green approaches for the one-step synthesis of fluorescent CDs can be obtained using natural or synthetic non-toxic precursors for specific biosensing purposes, e.g., wool for synthesizing CDs to detect glycophosphate, sodium fluoride-functionalized chitosan for preparing CDs to determine the retinoic acid content. The carbonaceous materials are possible to electrochemically, chemically, or physically dissect or cleave into small nano-sized fragments. The other approach involves either carbonization of small organic molecules or stepwise integration of small aromatic compounds.

4.3 CD Nanosensors

While monitoring the crop yield under varying stress conditions in addition to the applications of the pesticides at the time of final harvesting, it would be too late to take any preventive measures. One could only contemplate taking preventive measures in the next crop based on experience. It will be far more effective in taking preventive measures by detecting the onset of stress-related changes in the crop by detecting the signalling molecules using highly sensitive nanosensors. Similarly, instead of taking stock of the harmful effects of unused pesticides and their residues on overall agricultural production, it will be more manageable to take care of the damages caused using nanosensors for measuring the presence of harmful chemicals affecting environmental pollution and ultimately influencing human health through food chains operative in the ecosystem.

4.3.1 *Detecting Signalling Molecules in Plants: Abiotic and Biotic Stress*

The phenomenon of global warming, which is causing the resulting climate changes experienced in the current situation, is found to be influencing the occurrence of plant-stress-related events, which is causing the yield to suffer more and more. In this context, remote sensing technology has evolved to monitor and control plant stress-related problems to take care of the problems causing reductions in crop yield and quality. Traditional remote sensing methods can only detect changes in plant phenotype and physiological characteristics during stress, which are primarily accumulative in nature and lack the ability to detect changes in chemical signalling molecules in plants at the onset of the stress. Although fluorescent dyes could be used to label signalling molecules, they have other problems with photobleaching that do not support long-term monitoring as such. In contrast, the nanosensors are far better at detecting the presence of chemical signalling molecules in plants during long-term and real-time tracking. Accordingly, the nanosensors can be used to monitor crop maturity and health, besides detecting fertilizers, pesticides, and moisture in the soil. It becomes thus much easier to take quick decisions necessary for improving agricultural production. Emerging optical nanosensors are capable of assessing the chemical signals inside the plants, providing enough time for imaging devices to detect the signals, and indicating the need to improve plant growth conditions (Zhu et al. 2022).

Carbon nanotube (CNT)-based sensors have been reported widely to detect signalling molecules like H_2O_2, Ca^{2+}, and NO in plants under abiotic and biotic stresses, with the advantages of good fluorescence stability, long life, and emitting fluorescence in the relatively transparent near-infrared emission spectrum region of living tissues. For instance, the CNT-emitted NIR fluorescence spectra are quenched by H_2O_2 within the physiological range of 10–100 μM in plants, which can be deployed to remotely report plant stress conditions without causing mechanical leaf damage. Due to the sensitivity of 10 μM H_2O_2, hemin complexed-DNA aptamer-conjugated SWCNTs were thus developed for real-time monitoring of H_2O_2content in the leaves under UV, high light, and Flg22 stress. This determined the profile of H_2O_2 content in the leaf, which varied due to the stresses, indicating that hemin-complexed-DNA aptamer-conjugated SWCNTs may be a good candidate for early stress detection in plants. In addition, SWCNTs can be combined with a haemoglobin (Hb)-modified carbon fibre ultramicroelectrode (CFUME) to construct an *in vivo* H_2O_2 sensor based on direct electron transfer. These Hb/SWCNTs/CFUME sensors could detect a dynamic range of up to 0.405 mM H_2O_2 with a low detection limit of 4 μM H_2O_2 at a low working potential of −0.1 V. For monitoring plant diseases like Huanglong in citrus, SWCNTs functionalized with anti-SDE1 polyclonal antibodies could detect the SDE1 secreted protein, a biomarker of Huanglong disease, with high sensitivity and selectivity. This could overcome the traditional Huanglong disease nucleic acid problem and other problems of low diagnostic accuracy. Single-stranded DNA conjugated SWCNTs were used to detect volatile marker molecules such as ethylhexanol, linalool, tetradecene, and phenylacetaldehyde of Huanglong disease, which are lethal to citrusfruis, to realize the monitoring of the asymptomatic stage of the disease. Moreover, 6-ssDNA-modified SWCNTs could detect 4-ethylphenol, a major volatile organic compound released by *Phytophthora cactorum*-infected strawberries, within a few minutes via an electrostatic gating mechanism, with a wide concentration range and high detection sensitivity and recovery.

Such CNT-based nanosensors are noted as useful in monitoring stress signalling molecules via early detection of the plant experiencing the stress. However, to date, these studies have mostly been carried out at the laboratory scale, and the proposed approaches have not yet been conducted under actual agricultural conditions. The performance of these CNT-based nanosensors are affected by the factors like temperature (heat, cold, and freezing), wind, rain, light (high light, shading, and UV), and aerial pollutants and is still not known in detail. It is also necessary to conduct comprehensive research and development on the connection between nanosensor-based sensing, plant stress detection, data storage and analysis, and the agricultural equipment required therein. Moreover, the current studies of nanosensors deployed in stress detection are mainly focused on the detection of stress-related plant signalling molecules, with lesser attention paid to the monitoring of plant nutrient deficiencies, which are also important limiting factors of agricultural production (Zhu et al. 2022).

While applying pesticides for crop protection, the release of many compounds into the environment and ecosystem is practically unavoidable under practical conditions. The remaining pesticides and their residues are invariably found to cause serious problems in food safety, ecological pollution, and human health hazards. Traditional pesticide detection techniques like chromatography, electrophoresis, and mass spectrometry are in general not suitable for the real-time and long-term detection of small amounts of pesticide residues, and especially not for performing on-site detections. Despite their high sensitivity and accuracy, they are not easy to use, as the processes involved are time-consuming and require complicated sample preparations and expensive instrumentation. Using nanosensors for rapid analysis of pesticide residues appears to be a much-favoured option compared to the other techniques. Compared to the traditional methods, nanosensors have the additional advantages of being simple, having higher sensitivity, and having better selectivity during direct on-site detection.

CNM-nano sensors with autofluorescence have been widely studied for pesticide detection. The autofluorescence of carbon dots (CDs) is quenched in the presence of pesticides. For instance, residues like diazinon, glyphosate, and semi-carbazide in cherry tomatoes are detected by fluorescent sensors with detection limits as low as 0.25, 0.5, and 2 ng/mL, respectively, which are the lowest detection limits of typical detection methods like chromatography. Malathion, a broad-spectrum organophosphorus insecticide, can be detected by CDs and gold-based NP-based nanosensors. The Au-NPs quench the fluorescence of the CDs based on the principle of fluorescence resonance energy transfer, but the fluorescence is detected due to the presence of malathion. Because of the increasing intensity and the changes in colour from red to blue, it is possible to detect malathion quickly through simple visual observation. Another scheme for detection of the pesticide residues is graphene QDs widely used in biological imaging and optical sensing due to its lower cytotoxicity and excellent biocompatibility. For example, nitrogen-doped graphene QDs can be used to prepare strip sensors for pesticide detection. Using a uniformly imprinted polymer film on the surface of a nitrogen-doped graphene QD-strip, thiacloprid has been specifically recognized and detected, with a detection limit of 0.1–4 mg/L. In addition to graphene QDs, graphene as such has also been used to prepare sensors to detect pesticides. For example, by inhibiting the peroxidase activity of graphene, aromatic pesticides were detected by graphene doped with heteroatoms, which showed the changes in the colour reaction of hydrogen peroxide and tetramethylbenzidine dihydrochloride. It is therefore proposed that different element-doped graphene sensors are in principle feasible to design for detecting the different aromatic pesticides. Polydopamine-functionalized graphene nanozymes were found effective in detecting triazine pesticides by changing the activity of peroxidase through pesticides.

Based on these observations, it is concluded that CNM-based nanosensors are appropriate candidates for monitoring pesticide residues for the benefit of agricultural production. However, it is imperative to develop environmentally friendly and biocompatible CNM-based nanosensors to detect single or mixed pesticide residues in field applications. CNMs have been used to detect heavy metals (Zhu et al. 2022).

4.3.2 CNMs: Nanocarriers for Agrochemicals and Genetic Materials

Utilization of fertilizers and pesticides is essential for agricultural production. However, the utilization rates of most of the chemical fertilizers in plants are <50%, which not only limits agricultural production but also results in environmental pollution. Most of the fertilizers being water-soluble, are easily subjected to leaching in the soil, causing subsequent environmental pollution. Because of the possibility of multiple applications, such unavoidable fertiliser losses ultimately raise the overall costs. Compared to their conventional counterparts, the efficacy gain of nanofertilizers is 18–29% higher. NPs containing one or more elements required for plant growth can be directly used as nanofertilizers. As a carrier to enable slow and/or targeted delivery of nutrients into plants, NPs can also be used as nanofertilizers.

CNMs are excellent fertilizer carriers due to their stable molecular structure, good biocompatibility, lesser toxicity, and uniform dispersion in application. GOs are good carriers of trace elements. The oxygen-containing groups on its surface easily cause electrostatic adsorption of the trace elements and have two-phase release characteristics, showing quick release at the early stage followed by slow and continuous release subsequently. Indeed, GO sheets can deliver Zn and Cu elements more efficiently in wheat than zinc or copper salts. The negatively charged copper/CNF are also being used as a slow-release carrier of micronutrient copper. Combined CNF-Cu transfers from roots to shoots through xylem and slowly releases copper in the plants, resulting in significantly improved water absorption, germination rates, root/shoot ratio, and protein content of chicory. Negatively charged fullerenol has been explored as a leaf slow-release fertilizer to promote the absorption of Fe^{2+} in cucumber leaves and to take care of the iron deficiency. Addition of nanocarbon to slow-release fertilizer can significantly improve rice yield and nitrogen use efficiency, and reduce nitrogen loss, indicating that nanocarbon can be used as an environment-friendly slow-release fertilizer coating material (Zhu et al. 2022). These results show that CNMs can be good carriers to deliver nutrients to plants or improve the effect of fertilizers.

Based on past experiences, there are still public concerns regarding the biosafety and pollution issues of traditional pesticides due to their easy leaching, volatilization, and loss properties. Using pesticides in excess to protect the crops has been at the root of many problems like plant disease resistance, destruction of soil biodiversity, and adverse effects on human health and the environment. Therefore, more efficient and environmentally friendly solutions regarding the use of pesticides are necessary to evolve in the future. Nano-pesticides (including nano-insecticides, nano-herbicides, and nano-fungicides) are known to reduce volatilization and degradation, improve utilization efficiency, and are bound to reduce the overall use of the pesticides and alleviate environmental pollution to a significant extent. CNMs, being efficient adsorbers of harmful organic matter to reduce the solubility and bioavailability of organic matter, are also extremely good as pesticide nanocarriers to improve the utilization efficiency of pesticides. Currently, due to the unique physical and chemical properties mentioned above, GO is one of the most widely used pesticide carriers in the field of nanopesticides. For example, Reduced graphene oxide (rGO) has the advantages of high pesticide adsorption capacity, low toxicity, good antibacterial properties, insensitivity to pH value change, and the ability to be reused. GO loaded with red spider insecticide are completely adsorbed on the surface of the red spider, producing a strong toxic effect. GO can carry different types of pesticides through surface modification. Water-soluble polydopamine-modified GO was reported to alleviate the issue of easy loss of water-soluble pesticides by enabling controlled release of pesticides, enhancing the adhesion between pesticides and plants, and thus improving utilization efficiency. Hydrophobic pesticides are reported to be loaded by polylactic acid-modified GO. However, GO also has the disadvantage of having lower stability in acidic solutions. It is possible to control the delivery and release of emamectin benzoate with the help of a nano-biochar carrier of emamectin benzoate involving carboxymethyl chitosan as the pH-responsive switch. This resulted in improved water solubility, dispersion stability, and UV resistance of the delivery system, ensuring its long-term control of pests. In addition to GO, CNTs have also been used as a sustained-release system for pesticides. For example, MWCNTs grafted with polycitric acid (PCA) were reported to deliver zineb, an antifungal pesticide. Compared with zineb in bulk, the novel CNT-PCA-zineb hybrid material exhibited improved water solubility, higher stability, and stronger toxicity to *Alternaria*.

CNMs are promising candidates for delivering agrochemicals to plants more efficiently than conventional fertilizers and pesticides. However, it is necessary to better understand the plants' responses to exposure. The addition of CNMs was noted to increase the complexity of the agroecosystem, whether they represent a new pollutant or a new opportunity. In considering the future of nano-agrochemicals, it is necessary to fully consider the views in many fields of science, industry, and regulations so that the agrochemicals sector can make use of nanotechnology and reduce its negative impact on human beings and the environment as much as possible (Zhu et al. 2022).

To feed around 10 billion people in 2050, the breeding of stress-tolerant species is gaining more and more significance for ensuring the food supply in the future. At present, the commonly used transformation methods in plants, such as gene gun bombardment and *Agrobacterium*-mediated transformation, have problems such as restriction of the host type, reduction of transformation efficiency by the cell wall, damage to the plant tissue, and pathogenicity. In recent developments, nano-enabled transgenic studies have shown convincing progress. Out of the nanomaterials used, carbon nanotubes and carbon dots are important. Nanomaterials can deliver nucleic acids faster compared to *Agrobacterium*-mediated gene transformation. In addition, CDs can form relatively small complexes with nucleic acids, deliver the nucleic acids into cells, and facilitate tracking the complexes directly and conveniently. A PEI-modified CD-siRNA delivery system was reported to deliver and silence target genes in tobacco and tomato. In addition to CDs, CNTs are the most widely tested carriers for nano-enabled transgenic events. A method for delivering plasmid DNA with PEI-modified SWCNTs was found to effectively deliver plasmid DNA to wheat and cotton without transgene integration and showing highly expressed YFP protein. Further studies confirmed that SWCNTs could also deliver siRNA to mature plant leaves to achieve instantaneous gene silencing, with a gene knockout efficiency of around 95%. SWCNTs modified with chitosan can achieve chloroplast genetic transformation without external biological or chemical assistance, which is much cheaper and simpler compared to the conventional methods. Researchers also reported a system of SWCNTs and conventional cell-shuttle peptides to improve the efficiency of targeted peptide–plasmid transformation. CNMs are important in nano-enabled transgenic events and may be good candidates for targeted delivery of functional genetic materials. However, the plant transformation method based on CNMs is still at the embryonic stage and needs to be further studied in detail for its further applications.

4.3.3 CNMs: Light Converter for Augmenting Plant Photosynthesis

Plants are known to convert solar energy into chemical energy through chloroplast photosynthesis. However, the utilization rate of sunlight by chloroplasts is <10% and is limited to the visible spectral range (400–700 nm), mainly in the blue and red regions. The light from the UV and NIR regions does not participate in plant photosynthesis. Therefore, to expand the light spectrum for plant photosynthesis, developing high-performance light conversion materials with blue and red broadband emissions to make maximum use of solar energy might be considered a feasible approach for sustainable agriculture.

4.3.3.1 CNMs: Down-Conversion of Light

UV light (200–400 nm) gives rise to the generation of reactive oxygen species (ROS) in plants and affects agricultural production adversely. To increase the process of plant photosynthesis, various light-trapping nanomaterials have been explored to convert poorly absorbed UV light into highly absorbed visible light, improving the conversion efficiency of the light used by chloroplasts.

CNMs are used as down-conversion nanomaterials (DCNMs) to convert ultraviolet (UV) light into photosynthetic active radiation due to their stable emission and adjustability of the photoluminescence. Most of the CDs synthesized at present can emit blue fluorescence under UV excitation. Vinyl alcohol-encapsulated CDs are reported to convert UV to blue light and enhance the photosynthetic efficiency of lettuce. Amine-functionalized CDs can be strongly conjugated on chloroplast surfaces and promote photosynthesis by accelerating the conversion of solar energy. A new dual-wavelength luminescent CD was designed, exhibiting strong absorption of UV light and emitting light that exactly matched the chloroplast absorption spectrum (blue and red light). The adenosine triphosphate produced *in vitro* by the hybrid photosystem (chloroplasts coated with CDs) is 2.8 times that of the chloroplast itself. In the *in vivo* experiments, the electron transfer rate of the Rome lettuce leaves coated with CDs increased by 25% at the maximum. It should be noted that the impact of CDs on plant photosynthesis may be affected by the quantum yield. The chloroplasts using blue fluorescence re-emitted by CDs with medium QY (46.42%) could enhance photosynthesis, not low or high quantum yields. This amount to suggesting that the properties such as the emission intensity, quantum yield, and emission light spectrum of CNMs used for converting UV to visible light should be properly examined before they are designed for their uses in plants. Moreover, as a UV-visible light colour converter, CDs can also be applied to plastic films and LEDs in greenhouses to promote plant growth.

4.3.3.2 CNMs: Light Converters

About 52% of the solar spectrum in the NIR region is not used by the plants in photosynthesis, resulting in a significant waste of solar energy. The up-conversion nanomaterials (UCNMs) that can convert the NIR light into the visible region could possibly offer a better utilization of solar radiation by the plants. The conversion of NIR to visible light by UCNMs is a nonlinear process in which two or more low-energy photons from NIR light are converted into high-energy photons having a shorter wavelength and stronger energy via energy transfer, excited state absorption, or multiphoton absorption. CNMs work with up-conversion optical materials to promote NIR light conversion to enhance plant photosynthesis. Doping CDs with the up-conversion material NaYF4: Yb, Er can shift the green emission of NaYF4: Yb, Er to red light. Mung beans, for example, showed a significant increase in photosynthetic rate after being sprayed with NaYF4: Yb, Er@CD nanocomposite. It is also possible to use CNMs as up-conversion nanomaterials (UCNMs). However, the CNMs have rarely been applied as UCNMs for the improvement in plant photosynthetic efficiency.

Based on the observations reported above, it is worth examining the possibility of using CNMs to convert UV and NIR to visible light as a potential means to augment the process of plant photosynthesis. The leaf and/or root application of CNMs as light conversion materials for improving plant growth based on photosynthesis is still at the proof-of-concept stage based on the nonlinear interactions of the photons. For extending the concept of using CNMs as light converters in plants, a number of additional conditions must be fulfilled. For example, considerations of the biocompatibility, long-term safety, and toxicity of light-converting CNMs in plants besides the light conversion efficiency and the heat generation during CNM-enabled light conversion in plants become necessary to examine under practical conditions. After attaching NIR dyes to the surface of lanthanide-doped UCNPs, the up-conversion efficiency is more than 30,000 times higher than that of UCNPs alone. Moreover, some optical nanomaterials, such as carbon nanohorns (CNHs) and gold nanorods (Au NRs), generate local heat through photothermal conversion under NIR laser irradiation.

4.4 Chapter Summary

Applications of CNMs in agricultural production have been introduced in this chapter, with special emphasis on their applications as biosensors and carriers of nanofertilizers and nanopesticides, along with the possibility of using them for up- or down-conversion due to nonlinear photon interactions. Looking into these possibilities gives hope for exploring the carbon nanomaterials in improving the positive roles in future agricultural production, involving in increasing food production, and helping with the sustainability of agriculture. However, their effects may vary depending on factors such as plant species, CNM type, and dosages. Currently, the role of CNMs in agriculture is primarily limited to laboratory experiments. It is just the beginning, and it is necessary to conduct studies of large-scale quantitative field application data to allow for their future application in agriculture.

Further Reading

Mansuriya BD, Altintas Z (2021) Carbon dots: classification, properties, synthesis, characterization, and applications in health care – an updated review (2018–2021). Nanomaterials (Basel) 11. https://doi.org/10.3390/nano11102525

Mukherjee A, Majumdar S, Servin AD et al (2016) Carbon nanomaterials in agriculture: a critical review. Front Plant Sci 7. https://doi.org/10.3389/fpls.2016.00172

Zhu L, Chen L, Gu J et al (2022) Carbon-based nanomaterials for sustainable agriculture: their application as light converters, nanosensors, and delivery tools. Plants (Basel) 11. https://doi.org/10.3390/plants11040511

Chapter 5
Nanofertilizers in Agriculture

5.1 Introduction

Plants grow and survive by taking adequate nutrients from the soil. Loss of nutrients due to various reasons like leaching, volatilization, erosion, and uptake by the plants makes it necessary to compensate by using multiple-element fertilizers. The nutrient use efficiency (NUE) of plants lies in the range of 30–35%, 18–20%, and 35–40% for N, P, and K, respectively, in the case of conventional fertilizers. The use of biofertilizers in combination with or in place of chemical fertilizers has not been found very effective in this context. Currently, nano-fertilizers (NFs) are showing promise to improve the NUE and, eventually, the crop yield. The delivery of elements like P and Zn is generally very poorly bioavailable and reduces the loss of runaway nutrients such as nitrate. Nano-fertilizers are provided mainly as the encapsulation or coating of nutrients with nanoemulsions and nanoparticles (NPs), respectively. Such NFs serve as nutrient nanocarriers like nanoclays, hydroxyapatite NPs, polymeric NPs, carbon-based nanomaterials (CNMs), mesoporous silica, and miscellaneous materials. Nanoscale materials, including aptamer derivatives, carbon nanotubes, quantum dots, and others, are also being explored in the agriculture sector as nano-biosensors to indicate the presence of microbes, contaminants, pollutants, toxins, pH levels, nutrient levels, and moisture content. In general, all NFs provide a slow, steady, and time-dependent release of essential nutrients to ensure their delivery to the plant in a balanced and need-based manner. NFs can improve the NUE by about 3× and improve crop productivity by promoting seed germination, seedling growth, nitrogen metabolism, photosynthetic activity, protein synthesis, antioxidant defense, and other activities. Some limitations and adverse effects of NFs have also been reported; however, these are overcome by proper standardization of NF dose and selection of the right kind of NM for the test crop. The efficacy of the NF depends on its capacity for ionization and successful delivery of nutrients to the sink, which are modulated by the chemical composition of NMs, their

A. Nazeer et al., *Targeted Delivery of Nanopesticides and Nanofertilizers in Sustainable Agricultural Farming*, Nanotechnology in the Life Sciences,
https://doi.org/10.1007/978-3-031-41333-9_5

concentration and aggregation state, the metabolic potential of plant species, and the local environmental conditions. It is critical to determine whether NFs are fully transformed into ionic forms in the plant and then incorporated into proteins and various metabolites, or whether some of their parts remain unused and reach consumers via the food chain (Iqbal et al. 2019).

Nanoscale materials species of conventional fertilizers, also called nanofertilizers, provide nutrients and phosphates that, when added to the soil, offer enhanced delivery to improve their uptake and performance. Even though nanofertilizers and NM-enhanced fertilizers are very promising for agriculture, the use of nanofertilizers is yet to be explored in this area of research. In this context, hydroxyapatite-NPs were noted to enhance the soybean growth rate and seed yield by 33% and 20%, respectively, compared to a regular phosphorus fertilizer. In addition, slower releases of nanofertilizers are expected to contribute to maintaining soil fertility by reducing the transport of these nutrients into runoff or groundwater and decreasing the risks of environmental pollution and toxicity due to their overapplication, as reported in several reviews.

Oxide-NPs of Fe, Zn, Ti, and Cu have been applied directly as nanofertilizers in soil via irrigation or foliar application routes in plants such as the mung bean, cucumber, and rape. Similarly, soil supplements based on MWCNTs were found to increase the number of flowers and fruits in tomato plants, likely through the activation of genes and proteins essential for plant growth and development. Despite this evidence, the use of nanofertilizers is still not considered reliable. Accumulation in treated soils may pose a threat to soil microbial communities such as small invertebrates, bacteria, and fungi. Such impacts on the agro-ecosystem discourage the use of metallic NPs in agriculture.

Only recently, natural polymer-based chitosan NPs were used for controlled release of nitrogen, phosphorus, and potassium in wheat via foliar application, implying that organic-NPs are more environmentally friendly. However, their effective advantages for nutrient supply over traditional fertilization methods need more robust evidence through elaborate studies.

5.2 Organic Fertilizers

Organic farming is, of late, being promoted to restore soil health and fertility through the addition of organic matter. This is a common practise among farmers because it improves physical, chemical, and biological soil properties in addition to supplying plants with nutrients. This awareness is coming to the farmers, who should better return to traditional methods using crop residues and animal waste as manure. Some of the salient features of organic manure are summarised here.

Crop residues after harvesting are useful against soil erosion by improving the soil structure, increasing the organic matter contents, reducing evaporation, and fixing CO_2, but are usually disposed of in landfills or burned, aggravating air pollution due to emissions of particulate materials and CO_2. These residues, when used as green manure, can provide nitrogen to the soil through nitrogen fixation, increasing the supply of nitrogen to subsequent crops and reducing chemical fertilizer applications. Using green manure from legumes and non-legumes, independently and in combination, has been reported to improve the yield of various crops by improving the chemical, biochemical, and microbiological characteristics of the soil. Application of green manure has been found to improve bacterial communities by increasing biodiversity via decomposition. Certain green manures have been found useful in controlling weeds and nematodes and acting as biofumigants (Durán-Lara et al. 2020).

Animal manures are used in crops and grasslands, leading to a relevant reduction in the use of N-fertilizer. The availability of animal waste is anticipated to be popular in developing countries. Organic manures of animal origin include poultry, bovine, and sheep manure, among others. The availability and use of such manures for crop production depend on the geographical area, manure price, extent of production, and related management. The influence of the application of animal manures in reducing the toxicity of soils contaminated with heavy metals is finding applications as an alternative to phytoremediation.

Uric acid is the source of N_2, which is an inexpensive and commercially available fertilizer at large scales. However, the application of untreated bird manure might be risky as it harbours bacteria like *Escherichia coli*, *Listeria monocytogenes*, *Salmonella* spp., and other pathogens that might affect soil, irrigation water, and plants. Poultry manure is a very good source of nutrients that can be integrated into most of the fertilizer applications.

Cattle manure is another fertilizer supplement used in horticulture. Compared to mineral N-fertilizers, cattle manure is less effective for crops, but it increases soil organic carbon (SOC) and soil N, helping long-term soil fertility maintenance. Sheep manure, used as an organic fertilizer for improving the soil structure and causing enhanced fertility, is promising for improving soil structure, microbial activity, water use efficiency, and crop performance. The application of sheep manure to contaminated soil could decrease Cd, Pb, and Zn concentrations in plant tissues, reducing metal toxicity and metal uptake by plants. The animal manures, which contain a large volume of organic material, can be used for composting, but occasionally they might trigger environmental problems such as groundwater pollution by phosphorus and heavy metal contamination from the animal feed. These heavy metals in soil, water, and plants can bioaccumulate in the environment and subsequently in humans.

Because urine has urea/ammonium ratios comparable to synthetic fertilizers, it has been used to fertilize cucumbers, corn, cabbage, wheat, and tomatoes. Human urine used in cabbage cultivation in comparison to industrial fertilizer achieved the same fertilizer value as industrial fertilizers. Using urine could, for instance, increase the yield of red beets.

Compost, vermicompost, and biochar are important composting techniques for solid organic waste, which is a process that involves the bio-oxidative decomposition of organic matters. Composting produces high-quality, effective products for application in agriculture due to its cost-effectiveness, ease of operation, and environmental friendliness. The composting of cattle manure has a variety of agricultural benefits, such as decreasing the mass and water content, inhibiting pathogens, killing weed seeds, and producing stable and spreadable organic matter. Within the composting processes, greenhouse gases are produced, including nitrous oxide (N_2O), methane (CH_4), and ammonia (NH_3). Greenhouse gas emissions from compost can be reduced using chemical additives like salts of PO_4^{3-}, Mg^{2+}, superphosphate, and gypsum that assist chemical reactions in the compost substrate. Physical additives like biochar, zeolites, bentonites, sand, and soil are found adsorbing or changing the physical factors of the compost, and microbials like ammonia-oxidizing bacteria affect the renewal processes. Vermicomposting is another strategy wherein certain species of earthworms are used to accelerate the process of degradation, resulting in manure. This technique is faster because the organic materials, while passing through the earthworm gut, encounter microbial activity and plant-growth regulators in addition to being fortified with pest-repellent elements. In addition, it serves to reduce the concentrations and availability of heavy metals (Durán-Lara et al. 2020).

5.3 Nanofertilizers for Agriculture

Nanofertilizer (NF) is a material that is either a plant nutrient itself in the form of micro- or macro-nutrients or the carrier of some chosen plant nutrient. Nutrients encapsulated or covered by nanomaterials are also called NFs. They are developed from synthetic substances in the modified forms of synthetic fertilizers or green synthesized forms from different parts of plants through various chemical, mechanical, or biological methods using nanotechnology. The top-down syntheses of NPs involve the breaking down of bulk material into nano-sized material species, compared to the bottom-up approach, in which they are synthesized from atoms, molecules, and smaller monomers. Green synthesis is a non-toxic and environment-friendly method because it uses bio-organisms such as plants, fungi, and bacteria for the synthesis of NPs, and these micro-organisms work as both reducing and stabilizing agents without using hazardous chemicals.

NFs are primarily used for increasing soil fertility, increasing the bioavailability of plant nutrients, and improving product quality. Based on the nutrient requirements of the plants, NFs are usually categorized into macro-NFs, micro-NFs, and nanoparticulate fertilizers. NFs possess a characteristically slow and steady release of nutrients, both of which make them highly suitable for use in modern agriculture (Seleiman et al. 2021).

NFs are encapsulated with a nanomaterial that controls the release of nutrients according to the plant's requirements, and this causes an increase in the nutrient use efficiency (NUE) of plants. NFs release their nutrients in 40–50 days compared to synthetic fertilizers doing the same in 4–10 day durations, causing rapid loss of more than 70% of their N content after field application through leaching and volatilization, leaving less than 20% to be available for plants. Slow N release to plants was recently achieved by coating urea with hydroxyapatite-NPs. A nanohybrid of urea modified by hydroxyapatite was reported to release N as much as 12 times slower than synthetic urea in rice fields, and it could increase yields at only 50% the rate used with common urea. Similarly, synthetic P-fertilizers have low uptake efficiencies and high fixation rates in the soil compared to nano-formulations of P, reducing the nutrient losses via the direct internalization of crops. Using porous NMs, including chitosan and zeolites, has been found to improve uptake efficiency by controlling demand-based release and decreasing the loss of N. This depends on the method of application, i.e., foliar or soil amendments.

Despite numerous approaches proposed to improve P-uptake efficiency, success has been limited. However, higher solubility of phosphate minerals and increased uptake by plants were recorded after the application of ammonium zeolites. On the other hand, the use of P-enriched hydroxyapatite-NPs was found to considerably increase the plant height, shoot growth, and grain yield (18%) of soybeans compared to plants grown with synthetic P-fertilizers. Likewise, carbon-based NMs like graphene oxide films have the potential to extend the process of potassium nitrate release, thus decreasing leaching losses (Seleiman et al. 2021).

The metabolic processes influencing the mobilization of nutrients like P in plants are noted to be affected by nanoparticulate species, increasing the uptake efficiency of synthetic P, where Zn-NPs influence mobilization.

In advance formulations, biosensors could be attached to NFs to control the release of the nutrients and their bioavailability depending on the growth stage of the crop-a technology that is not applicable to synthetic fertilizers. Finally, the rates and costs of NF applications are typically lower than those of synthetic fertilizers, since NFs are required in relatively smaller quantities.

NFs have unique properties that make them more beneficial than synthetic fertilizers, particularly their ability to enter plants when applied as foliar or soil amendments due to their small particle size (100 nm), as well as being chemically reactive due to a large surface to volume ratio, a common feature of nanoscale material species, and thus NFs increase nutrient availability. NFs are soluble in water and can increase the dispersion of nutrients in soil and further increase their availability to plants, while synthetic fertilizers have a low solubility due to their large particle size and high adsorption and fixation to soil particles. Fertilizers are encapsulated in NPs to increase their uptake and availability to plants, as well as to reduce their bulk requirements, as noted in the higher availability of nutrients to plants throughout the growth period via the application of zeolite-based NFs. Furthermore, the slow and targeted nutrient release of NFs minimizes their toxicity to plants and decreases N losses via volatilization, leaching, fixation, and denitrification, as well as salt accumulation in soil.

5.4 Mechanisms of NF-Uptake

Knowing the pathways of uptake and movement of the NFs from the soil into the plants is valuable in deciding their suitable route of application. If NFs prefer xylem transport, then an irrigation system would be the best application, whereas phloem transport would necessitate exogenous application. The composition of NFs, the size of NPs, the physiology of the plants, and the pore diameters (5–20 nm) of the cell walls affect the transportation and accumulation of nutrients released from NFs in plants.

In foliar applications, NPs face the cuticular barrier before entering the plant tissues. The cuticle layer is a waxy coating on leaves that has two entry points, i.e., the lipophilic or cuticular pathway and the other as a hydrophilic or stomatal pathway. The lipophilic pathway is for nonpolar solutes that can enter the leaves through diffusion, while the hydrophilic pathway is for polar solutes. Accordingly, NPs or their aggregates that are smaller than 4.8 nm in diameter can easily enter the cuticle through the cuticular pathway. NPs larger than 5.0 nm can enter plants by foliar application. Polar-NPs, on the other hand, can enter via a hydrophilic or stomatal pathway. The differences in leaf morphology and the number and size of stomata among plant species can affect the uptake of NPs applied through foliar spray. The morphological dimensions of stomata are about 25 μm m in length and 3-10 μm m in width. Nevertheless, as a result of the physiological function and unique geometric structure of stomata, the real size exclusion limit of a stomatal hole for NP diffusion is still unclear.

Following the stomatal pathway, NPs can travel long distances through a plant's vascular system after entering the leaf apoplast. Since the vascular systems of plants are unidirectional and noncirculatory, the nutrients or photosynthates moving towards shoots (xylem) or roots (phloem) do not come back to their original sites. Therefore, foliar-applied NPs only have the phloem system option for uptake and translocation from leaves to roots. Foliar-applied micronutrient oxide NPs (24–47 nm) on watermelon penetrated the leaves by following the stomatal pathway and reached the watermelon roots through sieve tube phloem. However, after the foliar spray of Cu-NPs onto lettuce, 97–99% of the Cu-NPs were confined to the leaves of the lettuce, while only 1–3% were found in the root tissues. Soil-applied CuO-NPs (20–40 nm) translocated from the roots of maize to the shoots through the xylem, and then they moved back to the roots through the phloem. Such findings supported the concept that NPs can circulate within a plant both through the xylem and the phloem (Seleiman et al. 2021).

Multiple factors like plant morphology, growth stage, exposure conditions, particle size, and rhizosphere processes affect the root uptake of NPs. SiO_2-NPs (50 and 200 nm) were found entering the roots of *Arabidopsis thaliana*. Conversely, TiO_2-NPs (36–140 nm) remained in the root parenchyma of wheat without reaching the vascular system, while NPs with a diameter of <36 nm were stored in the roots of the wheat and then translocated to the rest of the plant.

The roots of *A. thaliana* produced a mucilage that facilitated the taking up of positively charged gold nanoparticles (12 nm) by the roots, while the same-sized negatively charged gold nanoparticles did not enter the root tissues. On the other hand, different plant species have shown different uptake capabilities for NPs, probably because of the variations in physiological and metabolic functions. The roots of wild *Azolla caroliniana* absorbed both 4 and 18 nm gold NPs, and *Myriophyllum simulans* absorbed only 4 nm gold NPs, while *Egeria densa* did not uptake gold NPs of any diameter. Likewise, the roots of tobacco plants did uptake gold nanoparticles ranging from 10 to 50 nm in diameter, whereas wheat roots did not do so.

NPs applied to soil are initially adsorbed on the surface of roots and then cross several barriers to reach a plant's vascular system. The first barrier is the root cuticle layer, which has a similar composition to that of the leaf cuticle layer. NPs cross the root surface cuticle and reach the root epidermis with the choice of following the apoplastic or symplastic pathways. In the apoplastic pathway, NPs first enter the cell wall pores and then move into intercellular spaces. However, the pore diameter of the cell wall, typically ranging between 5 and 20 nm, restricts the passage of NPs through the apoplastic pathway to particles less than 20 nm in diameter. Despite this, it might be possible for NPs to induce the destruction of the cell, and this would extend the size of pores. Additionally, it has been demonstrated that NPs can reach intercellular space in roots that have been exposed to diseases and herbivory by insects and soil microorganisms. Nevertheless, the main barrier in the apoplastic pathway is the Casparian strip around the vascular system, which prevents the direct entry of NPs to the vascular cylinder, although research has shown that ZnO-NPs (30 nm) were able to enter from the lateral root junction of maize, reaching the vascular system. Another possible route is the symplastic pathway, where NPs move from one cell to another through plasmodesmata and reach the central cylinder. They can then move to the aboveground parts of the plant via the transpiration stream through the xylem.

Fertilizers are indispensable for growth and for enhancing the yield and quality of crops since they can provide the required nutrients for plant development. Macronutrients such as N, P, and K are needed by plants in large quantities. Since most of these nutrients are not efficiently taken up by the plants, farmers prefer to use high fertilizer doses to partially supplement their low NUE values, which results in detrimental impacts on soil, water, and the overall environment. The use of NFs can increase the NUE of fertilizers, enhance crop yield and quality, and decrease the negative effects of synthetic fertilizers in the context of more sustainable agriculture. Nano-enabled fertilizers precisely release nutrients in the root zone of plants by preventing rapid changes in the chemical composition of the nutrients in the soil, which, in turn, reduces nutrient losses. Different types of NFs are produced depending on the material or carrier present in them, such as nanoparticulate species of hydroxyapatite, zeolite, mesoporous silica, nitrogen, copper, zinc, silica, carbon, and polymeric NPs (Seleiman et al. 2021).

5.5 Various Types of NFs

Nitrogen (N) is a basic component of several amino acids, proteins, DNA, ATP, chlorophylls, and cellular components. Most of the metabolic functions and regulatory pathways in plants depend on the adequate availability of N. Plants uptake N in the forms of NO^{-3} and NH^{+4}. One of the main constraints of synthetic N fertilizers is the high volatilization and leaching rates that occur during and immediately after their field application. To minimize these losses, N-based NFs could be utilized for the continuous supply of N at a slow release rate, as noted in zeo-urea nanofertilizer (N-NF) on maize plants, which reported a high nutrient uptake, vigorous plant growth and yield, and better grain quality compared to synthetic urea. N-NF applied to starflower reported, for instance, a significant improvement in plant growth that resulted in higher essential oil yields. Similarly, urea-modified zeolites were found to increase soybean seed yield. An N-NF developed by coating urea onto nanofilm was successfully used in *Brassica napus*. Both nano-N and chelated nano-N were effective in increasing the yield of a potato crop and decreasing nitrate leaching. Recently, NPK-coated NFs applied to greenhouse-grown coffee seedlings resulted in increased nutrient uptake and growth of coffee plants as measured by an increase in the number of leaves and photosynthetic plant area. The nitrogen in the form of NFs is highly recommended for its slow release of nitrogen, reduced volatilization and leaching rates, high nutrient uptake, and improved growth and productivity of the crops.

Phosphorus (P) is the second vital nutrient for optimum plant growth, as it is an integral part of energy transfer molecules like ATP, ADP, phospholipids, and sugar phosphate, and it influences processes like photosynthesis, respiration, and the biosynthesis of DNA. Various factors of plant productivity, such as root and shoot length, plant vigour, resistance to diseases, number of reproductive buds, yield, and quality, are strongly influenced by P. However, P in synthetic fertilizers is poorly available due to its slow releasing time and high fixation in soils in comparison to NFs, which gradually deliver P for up to 40–50 days following their application, while common P synthetic fertilizers deliver all the nutrients within 8–10 days post-application. Therefore, it has been suggested that the use of NFs or slow-release materials like zeolites may have the potential to increase the NUE of P for several field crops. A biosafe nanofertilizer was found to increase fresh and dry biomass, fruit yield, and quality by several times, in addition to leading to a high NUE. Likewise, the application of a nanosized hydroxyapatite to a soybean crop enhanced its growth, resulting in a seed yield that was 20.4% higher than that achieved with a synthetic P fertilizer. A significant enhancement in the growth and antioxidant contents was reported in *Adansonia digitata* plants treated with the foliar application of HA-NPs. P applied in NF form is a suitable option, particularly in smart agriculture, because it has a slow release material over a long period and can consequently reduce the leaching of P into groundwater and enhance the productivity and quality of crops.

Potassium (K) is the third important macronutrient, playing a vital regulatory role in all the physiochemical functions of plants to sustain normal growth and development. Potassium is involved in plant stomatal opening, photosynthesis, the translocation of photosynthates, protein synthesis, ionic balance, water relations, and the activation of more than 60 enzymes. Plants with an adequate quantity of potassium are more resistant to abiotic stresses such as water stress and high/low temperatures. On the other hand, potassium deficiency affects root shoot growth, the number of seeds inside fruits, size, shape, colour, taste, and the yield of crops. The maximum use efficiency of a potassium fertilizer is typically in the range of 30%

and 50%, indicating that up to 50–70% of the applied potassium fertilizer can be lost, causing substantial economic losses and deleterious effects on soil health and water quality. A nano-potassium fertilizer formulation was reported with slow K-release rate. The authors concluded that application of a nano-potassium fertilizer could reduce K losses in soil while sustaining the potassium supply to crops over a longer period of time. Potassium-loaded zeolites NFs increased the yield, harvest index, potassium concentration, and chlorophyll content in hot pepper. Similarly, a nano-potassium fertilizer via foliar application improved the growth, biomass, and quality of *Cucurbita pepo*. Using potassium-NFs can protect soil health and improve water quality by reducing K losses into soil and subsequently enhancing the physiological and yield traits (Seleiman et al. 2021).

Though required in considerably smaller quantities than macronutrients, micronutrients are also vital for maximizing plant productivity and quality and for increasing plant tolerance against multiple stresses. The synthesis of micronutrients by nanoparticulate structures may increase their solubility and bioavailability, aid in achieving a more uniform dispersal of these nutrients in the soil, and decrease the adsorption and fixation of micronutrients to the soil colloids.

Zn, being a structural component co-factor for various proteins and enzymes, influences plant growth significantly. Zn is also involved in the regulation of auxins, protein metabolism, the biosynthesis of carbohydrates, and the protection of a plant against pathogens and environmental stresses. ZnO-NPs are frequently used in modern agriculture as they are more efficient and cost-effective than synthetic Zn fertilizers and may be used for soil mixing or seed priming via foliar spray. However, traces of Zn can negatively affect plant growth by producing some metabolic alterations in plants if applied in high doses. The application of Zn-NFs can increase the germination, seedling growth, yield, and quality of crops. ZnO-NPs were found increasing the cabbage germination and tomato by improving protein, sugar contents, and antioxidants activities. ZnO-NPs applied to cucumber plants showed a higher uptake of ZnO than their synthetic bulk. However, a dose of 100 mg/kg of ZnO-NPs inhibited the growth traits of cucumber. The application of Z-NFs increased shoot growth, leaf area, dry weight, final yield, and protein contents in sunflower, pearl millet, rice, maize, sugarcane, and potato. Zn-NFs are considered the most used NFs in modern agriculture through foliar, soil mixing, and seed priming applications; they are also more cost-effective than synthetic Zn fertilizers in enhancing growth and improving the yield and quality of crops.

Iron (Fe) is an important nutrient involved in the synthesis of chlorophyll, DNA, chloroplast structure, respiration, and several metabolic pathways. Though plants need the element Fe in small quantities for their growth, its insufficiency or excess has detrimental effects on the physiological and metabolic functions of plants, thereby decreasing their yield. Iron availability in well-aerated soils is usually high. However, in these soils, Fe usually forms insoluble ferric compounds at neutral pH values, thus rendering it unavailable to plants. Therefore, Fe-enriched fertilizers could optimize the Fe supply to plants. Fe-NFs were found to increase germination and improve growth of different crops compared to control and/or synthetic Fe sources. Iron pyrite-NPs were reported to increase the growth of spinach. Better root growth in peanut plants treated with Fe-NPs was reported compared to non-treated plants under field conditions. Higher radical length during germination in green gram and higher fresh biomass were reported with Fe-NPs application (2–6 nm) compared to the control (ferrous sulphate; $FeSO_4$). Fe-NFs were shown to influence enhanced several growth parameters, chlorophyll and protein contents compared to untreated plants (Seleiman et al. 2021).

Manganese (Mn) is an essential micronutrient that is involved in N metabolism, photosynthesis, and the biosynthesis of fatty acids, ATP, and proteins, in spite of Mn being toxic to different plants. Manganese also helps plants cope with different stresses. Mn applications significantly improved the growth and yield of wheat, maize, sugarcane, soybeans, and common beans. Applications of Mn-NF on different crops showed that their treatments could enhance the root and shoot growth of mung bean by 52% and 38%, respectively, compared to a control treatment ($MnSO_4$; commercially available manganese salt). Mn treatments also enhanced the yield of eggplants by 22% and significantly increased the root length of lettuce compared to a control. However, there was no effect of Mn-NPs on the root length of white mustard, the seed germination of lettuce, or watermelon yield. Mn-NPs are attached with the chlorophyll binding protein (CP43) of photosystem II, and this results in an increased activity of the electron transport chain and, thus, the overall efficiency of the photosynthesis process. Consequently, plants fertilized with Mn-NPs have shown a higher rate of nitrogen assimilation and metabolism compared to their conventional bulk counterparts.

Copper (Cu) is a constituent of regulatory proteins that participate in photosynthesis and respiration in plants and is a cofactor of antioxidants such as superoxide dismutase and ascorbate oxidase. Copper deficiency leads to various disorders: necrosis, stunted growth, low numbers of seeds, grains, and fruits, and finally low crop yield. Soil organic matter content affects the availability of Cu, so the soil application of Cu-NPs may be beneficial due to their large specific surface area, high solubility, and reactivity. Recent field applications of a CuO-NPs nanofertilizer improved the germination and root growth of soybeans and chickpeas. Soybean seeds treated with nanocrystalline powders of Cu, Co, and Fr (40–60 nm) had 65%, 80%, and 80% germination rates, respectively, which were higher than the 55% germination rate in a control sample. Similarly, different concentrations of Cu-NPs increased the growth and yield of wheat due to improvements in leaf area, chlorophyll contents, number of grains per spike, and grain weight. Moreover, there was an improvement in flavonoid contents, sulphur assimilation, and the biosynthesis of proline and glutathione in *Arabidopsis thaliana* after the application of Cu-NPs with a dose of 5 mg/L. Conversely, Cu-NP application negatively affected the growth of water lettuce and decreased fruit firmness in cucumber plants. It seems that Cu-NFs can significantly and positively enhance biochemical and yield traits, but one must be careful with their rate of application.

Silicon (Si) has been ranked in between essential and nonessential elements for plants because it is not necessary for the completion of a plant's life cycle. However, under normal and stressful stress conditions, it provides certain benefits to some plants. Silicon is abundantly present in the earth's crust; however, soil Si uptake by plants only occurs in the form of monosilicic acid. Recently, attention has been given to Si due to its diverse role in plants against various stresses as noted that Si can improve the plant tolerance against heavy metal toxicity, as well as heat, water, and salinity stresses. The application of SiO_2 with organic fertilizers has the potential to improve overall plant productivity. The mesoporous structure of Si-NPs enables them to be suitable nanocarriers for various molecules that are beneficial in agricultural systems, as seen in their applications in the form of nanosensors and nanozeolites, which comprise the structure of Si-NPs for monitoring soil moisture and enhancing the water retention of soil. Si-NPs have the potential for specific plants that cannot survive without a suitable quantity of silicon or as nano-carriers to improve sustainable agriculture.

Boron (B) is an important micronutrient that is important in the elongation of pollen grains and tubes, formation of cellular walls, transfer of photosynthetic organisms from leaves to active sites, and increases in flower and fruit yields. B-NFs can improve plant growth and increase yield. Spraying 90–180 mg/L of B-NPs on mung bean crops reported higher number of pods per plant and a greater seed yield compared to a control. Likewise, spraying B-NPs at 20 ppm and nano-zinc at 200 ppm on olive trees yielded a maximum number of fruits with a high seed oil content. Similarly, a greater number of fruits and a higher yield were reported in pomegranate (*Punica granatum*) after the application of B-NF. Application of B-NF to an alfalfa crop grown on calcareous soil exhibited maximum yield with suitable forage quality (Seleiman et al. 2021).

5.6 Chapter Summary

Although encouraging NUE results have been found when using nanofertilizers, some limitations and adverse effects have also been reported. Most of the research has also only been carried out at a laboratory scale. Examples of shortcomings in the use of foliar fertilizers are that they require an available leaf area to be effective and may cause scorching or burning if the concentration of the spray is too high. They also require perfect timing for the application since climatic conditions affect effectiveness. The costs of multiple applications can be too high to be profitable. The standardization of the nanoformulations, the lack of size uniformity of the nanoparticles, and optimizing foliar applications of nanofertilizers are challenges that need to be addressed in future research.

Further research is required to better understand the role of different N forms in supplying these molecules (i.e., NO_3^-, NH_4^+) and their impact on NUE in pastures, including N transformation within the plant and their effect on reducing N losses to the environment. Additionally, the integration of this technology with the use of N cycling inhibitors may represent an opportunity by creating favoring synergetic effects. More information is required to understand whether nanofertilizers are fully transformed into ionic forms in the plant and later incorporated into proteins and different metabolites, or if some of them remain intact and reach consumers through the food chain.

A better understanding of the advantages of nanomaterials is needed; more and higher-quality data are required on material characterization, comprehensive comparisons with non-nano formulations, and field studies. Additionally, adapted nutrient fertilizer application strategies will need to be adjusted to take advantage of the benefits provided and to consider the inclusion of precision agriculture, such as drones fitted with cameras to gather multispectral images detecting N concentration in the grass within the paddock, avoiding over-application (e.g., urine patches), and repeated applications at low rates with a potential increase in costs. All these factors will need to be addressed in an integrated manner to account for potential downfalls and take full advantage of the opportunities and synergies the use of nanofertilizers may provide for the sustainable future of grassland production. The future development and adoption of these molecules as an answer for increasing food production with higher nutrient efficiency will need to balance economic and environmental costs of production with the potential reduction of environmental impact and yield increases. For this, a life cycle assessment (LCA) might offer an opportunity for an integrated analysis of its use, considering yield productivity, environmental implications, and impact on food chains. The need to validate the pros and cons of nanofertilizers under representative field grassland conditions to address the questions arising from stakeholders also remains a pending task before the widespread adoption of this technology.

Further Reading

Durán-Lara EF, Valderrama A, Marican A (2020) Natural organic compounds for application in organic farming. Agriculture 10. https://doi.org/10.3390/agriculture10020041

Iqbal M, Umar S, Mahmooduzzafar (2019) Nano-fertilization to enhance nutrient use efficiency and productivity of crop plants. In: Husen A, Iqbal M (eds) Nanomaterials and plant potential. Springer, Cham, pp 473–505. https://doi.org/10.1007/978-3-030-05569-1_19

Seleiman MF, Almutairi KF, Alotaibi M et al (2021) Nano-fertilization as an emerging fertilization technique: why can modern agriculture benefit from its use? Plants 10. https://doi.org/10.3390/plants10010002

Chapter 6
Pesticides and Crop Protection

6.1 Introduction

Before going into the further details of examining pesticides for their agricultural applications, it will be better to examine the proper definition of pesticides according to the FAO and UN.

Agriculture, forestry, aquaculture, food industry, processing, transportation, and storage of wood and other biological products are sectors that face a variety of pests like weeds, insects, birds, rodents, fungi, moulds, and the like, significantly affecting the quantity and quality of products. Certain species of pests can also damage buildings, installations, furniture, cloths, and others by spreading infectious diseases. Therefore, pest control is an activity needed to reduce or eliminate yield losses, maintain high product quality, and prevent other undesirable effects of the pests.

Pest control has been achieved by physical, chemical, and biological methods, and the choice of the particular method depends on conditions like the type of industry, species, amount and distribution of pests, and environmental conditions. The most effective methods of pest control are based on the use of chemicals referred to as "pesticides." The two largest industries using pesticides are agriculture and forestry. Adequate application of pesticides produces good results in terms of protecting products, improving quality, saving labour costs, and achieving a significant economic benefit.

The Food and Agriculture Organization of the United Nations (FAO) defines pesticides as follows:

Pesticide means a substance (or mixture of substances) of chemical or biological composition intended for repelling, destroying, or controlling any pest or regulating plant growth.

A. Nazeer et al., *Targeted Delivery of Nanopesticides and Nanofertilizers in Sustainable Agricultural Farming*, Nanotechnology in the Life Sciences,
https://doi.org/10.1007/978-3-031-41333-9_6

Pesticide is a more general term than plant protection product (PPP). Plant protection products are "pesticides" that protect crops or desirable or useful plants. They contain at least one active substance and have one of the following functions:

- protect plants or plant products against pests/diseases, before or after harvest;
- influence the life processes of plants (such as substances influencing their growth, excluding nutrients);
- preserve plant products, and
- destroy or prevent the growth of undesirable plants or parts of plants.

The term "pesticide" is often used interchangeably with "plant protection product," however, "pesticide" is a broader term that also covers non-plant or non-crop uses, for example, "biocides." Within EU legislation, there are separate legislative acts for PPP and biocides.

6.2 Classification of Pesticides

In general, groups of pesticides are classified and named according to the type of pest they control, as described in Table 6.1.

The growing world population is anticipated to impose a large burden on agriculture and its allied sectors in terms of meeting the rising food demands, which requires more inputs for crop production. Anthropogenic activities have affected our surroundings and have also had negative impacts on the environment and ecosystems, including reductions in agricultural areas due to construction, the explosion of nutrient mining, the degradation and contamination of water resources, the aggregation of xenobiotics in the soils, and the degeneration and deterioration of the quality, fertility, and efficiency of the soil, with implications for soil erosion and climate change. To overcome these challenges and meet the requirements for food and other supplies, the productivity and sustainability of agricultural practices must be improved by deploying novel and improved methods. Enhanced agricultural productivity can be achieved in many ways, such as by increasing crop yield by providing manure and organic-based treatments, including biopesticides, or by limiting yield loss due to extreme environmental conditions such as biotic and abiotic stresses.

Abiotic stresses are largely controlled by the use of biostimulants and bioeffectors. Biopesticides, which are pest management agents based on living microorganisms or natural products, offer better promises for controlling yield loss without compromising the quality of the product. The chemical pesticides used in crop protection to reduce the damage caused by pathogens and pests in agricultural fields pose long-term threats and risks to living beings due to their harmful side effects. They are known to cause cancer and foetal impairments, and they persist in the environment for many years. Furthermore, based on their potential applications and strong inhibitory activities against pests, these synthetic pesticides dominate the market, impacting product manufacture. Nutrient reduction and an increased

Table 6.1 Broad classification of pesticides

Organism	Uses	Reference
Bacteria	Insecticide (*Bacillus thuringiensis* var. *Kurstaki*; *B. thuringiensis* var. *tenebrionis*) kill caterpillars, Elm leaf beetle, and alfalfa weevil in the crops of vegetables, fruits, ornamentals, cereals, and potato. Fungicide (*Bacillus subtilis*) acts against *Botrytis* spp. existing in vegetables, fruits, and ornamentals.	Saberi et al. (2020), Kumar et al. (2021)
Fungi	Insecticide (*Bacillus thuringiensis* var. *kurstaki*; *B. thuringiensis* var. *tenebrionis* kill caterpillars, fungi (*Botrytis*), elm leaf beetle, and alfalfa weevil in the crops of vegetables, fruits, ornamentals, cereals, and potato. Fungicide (*Bacillus subtilis*) acts against *Botrytis* spp. existing in vegetables, fruits, and ornamentals. Herbicide (*Chondrostereum purpureum*) stops cut stumps of hardwood trees and shrubs in forestry. Nematicide (*Paecilomyces lilacinus*) controls nematodes in vegetables, soft fruit, citrus fruits, ornamentals plants, tobacco, and turf.	Saberi et al. (2020), McGuire and Northfield (2020), Moreno-Gavíra et al. (2020), Kumar et al. (2021)
Virus	Insecticide (*Cydiapomonella granulovirus*) treats codling moth in apples and pears.	Kadoić Balaško et al. (2020)
Oomycetes	Herbicide (*Phytophthora palmivora*) treat *Morenia orderatain* citrus crops.	
Neem	Insecticide (Azadirachtin) kills aphids, scale, thrips, whitefly, weevils, and leafhoppers in vegetables, fruits, herbs, and ornamental crops.	Chaudhary et al. (2017)
Plant extracts	Fungicide (giant knotweed extract) protects from powdery mildew, downy mildew, *Botrytis*, late blight, and citrus canker in ornamental and edible crops. Herbicide (plant essential oils) helps in eliminating ragwort, many arthropods in grassland. Nematicide (*Quillaja saponaria*) kills plant parasitic nematodes in vineyards, orchards, field crops, ornamentals, and turf.	Isman (2020), Guerra and Sepúlveda (2021)
Biopesticides	*Talaromyces flavus*, *Clitoria ternatea*, *Trichoderma harzianum*, *Bacillus thuringiensis* var. *Tenebrionis*, and *Lactobacillus casei* fermentation products are used in protecting plants like strawberry, cotton, gladiolus hybrids, alder leaf, and hazelnut, and other economically important plants and trees from *Glomerella cingulate*, *Colletotrichum acutatam*, *Helicoverpa* spp., *Fusarium oxysporum*, *Agelastica alni*, *Spodoptera litura*, *Helicoverpa armigera*, *Aphis gossypii*, *Xanthomonas fragariae*, *Spodoptera littoralis*, etc.	Kirk and Schafer (2015), Eski et al. (2017), Dubois et al. (2017), Pavela et al. (2017), El-Abbassi et al. (2017)

(continued)

Table 6.1 (continued)

Organism	Uses	Reference
Semiochemical	Citronellol attracts tetranychid mites in apples, cucurbits, grapes, hops, nuts, pears, stone fruit, nursery, and ornamental crops.	Mauchline et al. (2018), Mossa et al. (2017)
	Sex pheromone like (E,E)-8,10-dodecadien-1-ol attracts codling moth in apples and pears.	
Arbuscular Mycorrhizal Fungi	Mutual inhabitant in roots with species like *Fusarium verticillioides*; pathogens affecting below ground plant organs in *Zea mays*.	Olowe et al. (2018), Kumar et al. (2021)
Microalgae	Filamentous cyanobacterium (green algae) kill *Nostoc piscinale*, *Chlamydopodium fusiforme*, and *Chlorella vulgaris*.	Ranglová et al. (2021)
	Anabaena laxa and *Calothrix elenkiniiimprove* fungicidal activity in coriander, cumin, and fennel.	
Nanobiopesticides	Nanoparticulte silver kills *Alternaria alternata*, *A. Solani* in causing leaf blight and leaf spot diseases in tomato, pepper, and potato.	Narware et al. (2019), Rodrigues et al. (2019)
	Sargassum muticum NPs affect *Ariadne merione*, a Lepidopteran pest.	
	Caulerpa scalpelliformis and *Mesocyclops longisetus*-derived NPs kill *Culex quinquefasciatus*.	

disease incidence are quite common in crops grown on soils heavily subjected to chemical pesticides, and this is undesirable from an agricultural soil management standpoint for food and nutritional security. According to the Food and Agriculture Organization (FAO), UN, the top three leading pesticide-consuming countries are China, the USA, and Brazil. In addition, pesticide consumption in India drastically increased from 50,410 tonnes in 2016 to 58,160 tonnes in 2018.

The proportional use of pesticides in various crops provides some insight into the distributions. These are 67%, 50%, 46%, 43%, 28%, and 23% in fibre crops, fruits, vegetables, spices, oilseeds, and pulses, respectively. The FAO reported that from 2015 to 2018, the regional shares of global pesticide consumption were 52.2%, 32.4%, 11.8%, 2%, and 1.6% in Asia, the USA, Europe, Africa, and Oceania, respectively. Pesticide consumption per hectare was highest in China, followed by the United Kingdom, and lowest in India. Based on the data on chemical pesticide consumption in India, it is imperative to increase the use of biopesticides. Biopesticides are naturally occurring compounds or agents that are obtained from animals, plants, and microorganisms, such as bacteria, cyanobacteria, and microalgae, and are helpful in controlling agricultural pests and pathogens. According to the US Environmental Protection Agency, biopesticides are "derived from natural materials such as animals, plants, bacteria, and certain minerals." Products such as genes or metabolites from these biocontrol agents are used to prevent crop damage. The use of biopesticides is more advantageous than the use of their counterparts, traditional chemical pesticides, as they are eco-friendly and host-specific. The use and application of agro-based chemicals in the agricultural sector to protect crop plants from invading and infecting pests can be greatly improved by employing biopesticides.

6.3 Types of Biopesticides

Several types of biopesticides are classified in Table I in general, but one can particularly identify them according to their extraction sources and the type of molecule or compound used for their preparation, as described below.

6.3.1 Microbial Pesticides

Microorganisms like bacteria, fungi, and viruses fall into this category. The active molecules or compounds isolated from these organisms attack specific pest species or entomopathogenic nematodes. The bioinsecticides target insects that harm crops, while those that control weeds via microorganisms, such as fungi, are referred to as bioherbicides. Over the last decade, extensive study of microbial biopesticides has led to the development of a number of biopesticides for their entry into the market. The successful use of *Bacillus thuringiensis* (Bt) and some other microbial species facilitated finding newer strains with valuable toxins and virulence factors for industrial developments, and some of these have been translated into commercial products as well. The species of *Pseudomonas*, *Yersinia*, and *Chromobacterium* are bacterial entomopathogens, while fungi including *Beauveria*, *Metarhizium*, *Verticillium*, *Lecanicillium*, *Hirsutella*, and *Paecilomyces* have useful properties. Other important pesticide producers are baculoviruses, whose infectivity is associated with the crystalline occlusion bodies that are active against chewing insects. The baculoviral occlusion body is basically a virion, and combined with the Bt toxin, it produces recombinant baculovirus (ColorBtrus) for incorporating the Bt insecticidal Cry1Ac toxin protein to enhance the speed of action and pathogenicity with respect to its wild-type counterpart. Entomopathogenic nematodes used as biocontrol agents belong to species in the genera *Heterorhabditis* and *Steinernema*, associated with mutualistic symbiotic bacteria of the genera *Photorhabdus* and *Xenorhabdus* are safe for mammals, the environment, and nontarget organisms. Their commercial development as biocontrol agents has been convenient because of their ease in mass production, use of *in vivo* or *in vitro* techniques, and exemption from registration.

6.3.2 Biochemical Pesticides

Biochemical pesticides are used to control pests through nontoxic mechanisms, whereas chemical pesticides use synthetic molecules that directly kill the pests. Biochemical pesticides are further classified into different types depending upon whether they function in controlling infestations of insect pests by exploiting pheromones, plant extracts or oils, or natural insect growth regulators.

6.3.2.1 Insect Pheromones

These are chemicals produced by insects that can be used in integrated pest management to control the insects. These chemicals are effective in disrupting insect mating, thus reducing the number of insect progeny. The insects exploited in this process act as dispensers of pheromones that become confused due to the presence of pheromone fumes diffused in the surroundings. Insect pheromones are not true "insecticides," since they do not kill insects but influence their olfactory system to affect their behaviour. The insect antennae adsorb the pheromones, which then diffuse into the interior of the sensilla through micropores in the cuticle. Once inside, these are transferred through the hydrophilic sensillum to the chemosensory membranes by pheromone-binding proteins. Subsequently, the pheromone or pheromone-PBP complex interacts with a specific receptor protein that transduces the chemical signal into an electric signal by a second messenger system connected with neuronal machinery, as discussed in detail in the cited references.

6.3.2.2 Plant-Based Extracts and Essential Oils

Plant-based extracts and essential oils have emerged during the last several years as good alternatives to synthetic insecticides for insect pest control. These insecticides are naturally occurring as they are derived from plants and contain a range of bioactive chemicals. Depending on the physiological characteristics of the insect species as well as the type of plant, plant extracts and essential oils can act as repellents, attractants, or antifeedants against insects; they can also inhibit respiration, impede insect identification of host plants, inhibit oviposition, and reduce adult emergence via ovicidal and larvicidal effects. Well-known examples in this regard are neem and lemongrass oils, which are very common in global herbal markets. A comprehensive study showed that a combination of neem oil with entomopathogenic microorganisms like *Beauveria bassiana* was very successful against vegetable-sucking pests. However, it is critical to determine the dose of azadirachtin content in neem oil in order to avoid killing nontarget organisms. A similar strategy has to be chosen for the entomopathogenic fungi that are supported by complementary laboratory bioassays, station studies, and/or field experiments for effective management of the target pests without affecting nontarget insects. The marketability of essential oils is estimated at USD 700.00 million and a total world production of 45,000 tons, and industries in the US are able to bring essential oil-based pesticides to market in a short time period, as compared to the time taken in conventional pesticide launches.

6.3.2.3 Insect Growth Regulators

Insect growth regulators (IGRs) inhibit certain fundamental processes required for the survival of insects, thereby killing them. These compounds are highly selective and less toxic to nontarget organisms. Depending on the mode of action, IGRs have been recently grouped into chitin synthesis inhibitors (CSIs) and substances that interfere with the action of insect hormones (i.e., juvenile hormone analogues and ecdysteroids). IGRs can control many types of insects, like fleas, cockroaches, and mosquitoes, even though they are not so deadly for adult insects. Although low in toxicity to humans, they prevent reproduction, egg hatching, and moulting from one stage to the next in the young insects, while mixing them with other insecticides is able to kill even the adult insects.

6.3.2.4 GMO Products

These substances are produced through genetically modified organisms (GMOs) and incorporated into the plants, where they are used as a source of pesticidal compounds, also referred to as "plant-incorporated protectants" (PIPs). Cry proteins are by far the first-generation insecticidal PIPs that were introduced into the GM crops containing transgenes from the soil bacterium Bt. PIPs also demand the state of the research necessary for the ongoing environmental fate assessment of these molecules, primarily the RNAi-based PIPs.

6.4 Mode of Action of Biopesticides

Biopesticides affect microorganisms in a variety of ways, depending on their type and nature. A few mechanisms through which biopesticides attack or kill pathogens are listed as follows:

6.4.1 Microbial Biopesticides, Fungicides, and Bactericides

These biopesticides inhibit or disrupt the process of translation and thus protein synthesis in numerous ways, including through the binding of 50S ribosomes in prokaryotes to prevent the transfer of peptides and inhibit chain elongation (such as blasticidin). Sometimes they interfere with the binding of aminoacyl tRNA to 30S and 70S ribosomal subunit complexes and inhibit kasugamycin. In the case of streptomycin and mildiomycin, binding with the 30S ribosomal subunit causes abnormal protein synthesis and blocks the activity of peptidyltransferase, respectively. They also disrupt plasma membrane permeability and cause leakage of amino acids and electrolytes, causing cell death (such as natamycin), and can inhibit chitin synthase activity (polyoxins) and inhibit trehalase, preventing the formation of glucose (validamycin). Insecticides, upon reaching nerve endings, release gamma-aminobutyric acid (GABA), which causes GABA-gated Cl-ion channels to open, thus working by hyperpolarizing the nerve membrane potential and blocking the electrical nerve conduction (avermectins and emamectin). Polynactins facilitate the leakage of potassium ions from mitochondria. Herbicides inhibit phosphorylation in plants by blocking glutamine synthase, which causes an increase in ammonia (bilanafos).

6.4.2 Biochemical Pesticides

These pesticides are derived from plants. Plants have evolved and developed many compounds that can help combat pathogenic microorganisms during the course of infection and attack. These compounds include steroids, alkaloids, phenylpropanoids, phenolics, terpenoids, and nitrogenated compounds. For instance, nicotine was the first insecticide obtained from tobacco leaves in the seventeenth century that was used to kill plum beetles. Nicotine in tobacco is toxic to most herbivore insects, and pesticides derived from it have been regarded as "green pesticides" with high activity and low toxicity. The tobacco contained some useful ingredients, such as solanesol and nicotine, that exhibit potent inhibitory activity against *Staphylococcus aureus*, *Bacillus subtilis*, and *Micrococcus lysodeikticus*. Insecticides, such as azadirachtin and nicotine, function by either disrupting respiratory enzymes or inhibiting insect growth regulators, or by binding to sodium channels, while microbicides impair metabolic function, disrupt the integrity of the plasma membrane, and inhibit conidial formation.

6.4.3 GMO-Based Biopesticides

Using these formulations transfers the genes into a plant, which allows it to produce compounds like Bt toxin that combat pests. The delta endotoxins produced by the bacterium *B. thuringiensis* are broken down into smaller toxins in the insect gut by the action of proteases, which then bind to receptors in the midgut, causing cell expansion, rupture, and ion leakage, leading to cell death.

6.4.4 Biopesticides from Algae and Cyanobacteria

Microalgae offer an alternative technology to increase productivity in sustainable agriculture. A number of microalgae strains produce biologically active compounds that include antimicrobial compounds with the potential to act as biopesticides. The biomass extracts can be applied as an alternative to chemical pesticides since they can enhance plant growth and protect agricultural crops. The filamentous cyanobacterium *Nostoc piscinale* and two single-celled green algae, *Chlamydopodium fusiforme* and *Chlorella vulgaris*, are reported to have biopesticide activity against certain pathogens. Some important microalgae have been exploited for their beneficial biopesticide activity in the cultivation of spices.

The use of chemical insecticides can result in numerous undesirable effects, including the following:

- killing of beneficial or nontargeted organisms and sometimes resurgence
- rapid multiplication of secondary pests
- development of pesticide resistance
- contamination of the environment or ecosystem
- accumulation of pesticide residues in food materials
- imbalanced ecological processes, pollination, and harm to living beings
- carcinogenic and teratogenic effects in nature, and
- causing imbalances in hormone systems.

Out of several microorganisms deployed as biopesticides, microalgae, for example, are excellent sources with several advantages over traditional chemical pesticides. They produce a large number of compounds with stimulating activities, including biomass and compounds, which can be used in preparing biopesticides, thereby enhancing crop protection. Microalgae can be produced using wastewater, as they require nitrogen, phosphorus, carbon, and ammonium, which are abundant in wastewater. *Chlorella vulgaris* is generally used in the treatment of wastewater and is able to tolerate ammonium levels effectively. The efficacy of *C. vulgaris* was assayed against several phytopathogens like *Rhizoctonia solani*, *Fusarium oxysporum*, *Phytophthora capsica*, *Pythium ultimum*, *Clavibacter michiganensis*, *Xanthomonas campestris*, *Pseudomonas syringae*, and *Pectobacterium carotovorum*, while observing its antibacterial and antifungal activity, which were higher when cultivated in wastewater. It was found that rice fields sprayed with synthetic fertilisers left many detrimental effects on the environment and beneficial soil microflora, including decreased efficiency of fertiliser utilisation by promoting rice diseases, inhibition of microbiological nitrogen fixation, and increased nonpoint source pollution; importantly, they were also not cost-effective. Furthermore, in developing green rice, *Anabaena variabilis* could be a potent biofertilizer and biopesticide, as reported.

6.4.5 RNAi-Based Biopesticides

RNAi (RNA interference) technology is currently being used in the development of biopesticides with increased sensitivity to pests and pathogens. Many transgenic crops, like maize, soybeans, and cotton, have been developed with resistance against particular pests. Due to the limited consumption of genetically modified crops, RNAi can be used as an alternative to overcome this problem. Studies have demonstrated that transgenes significantly impact plant functions upon viral infection through an RNAi mechanism. A barley crop was reported to be completely resistant to the Barley yellow dwarf virus. The mechanism of RNAi includes the expression of transgene dsRNA, which induces virus resistance, and gene silencing in plants. Guide RNAs are formed as intermediaries, guiding target RNAs in their degradation. The use of RNA polymerase RDR6 was found to produce double-stranded RNA (dsRNA) from target transcripts in plants, leading to the formation of small interfering RNA (siRNA), which, in turn, showed silencing potential. The RNase III domain-containing enzyme responsible for dsRNA cleavage in *Drosophila* is known as Dicer and is also seen in plants and fungi. Following that, an RNA-induced silencing complex, a member of the Argonaute family, is used to mediate the cleavage of the target transcript, conferring host resistance. RNAi technology has been used to overcome the adverse effects of pests and pathogens. An RNAi method for oral application was developed using an artificial diet or transgenic maize against western corn rootworm to target V-ATPase subunits and alpha-tubulin. Similarly, investigations demonstrated the induction of growth defects in *Helicoverpa armigera*, the cotton bollworm, when given plant leaf material expressing a dsRNA specific to a cytochrome P450 gene. The first commercial, genetically modified variety showed the expression of dsRNA against an insect pest, as reported in 2017, when Monsanto and Dow approved SmartStax PRO maize containing dsRNA against the western corn rootworm Snf7 gene. Similarly, apple and potato-expressing dsRNAs were approved for regulation of endogenous gene expression for quality enhancement.

Apart from insects and viruses, the mechanism of RNAi-mediated silencing has been used to control other plant pests and pathogens, including *Agrobacterium*, powdery mildew, and root-knot nematodes. The US Environmental Protection Agency approved the first product called “SmartStax Pro” in June 2017 to help US farmers control rootworm, a corn pest that has developed resistance to several other pesticides as described elsewhere (Kumar et al. 2021).

6.4.6 Bacteria Based Biopesticide

Microorganisms and their product-based pesticides are highly effective, species-specific, and eco-friendly, leading to acceptance of their use in pest management strategies worldwide. There is enough scope for further development in their marketing and profitability in the manufacturing industry. The bacteria used for biopesticide actions can be divided into four categories, namely: crystalliferous spore formers like *Bacillus thuringiensis*; obligate pathogens like *B. popilliae*; pathogens like *Serratia marcescens*; and facultative pathogens like *Pseudomonas aeruginosa*. Of these, spore-forming bacteria are the most sought-after for commercial use. The commonly used bacteria, *B. thuringiensis* and *B. sphaericus*, are highly specific, safe, and effective organisms for insect control. The Cry family of proteins is produced by *B. thuringiensis* in the parasporal crystals and encoded by the cry genes. The Cry proteins are globular molecules with three structural domains connected by linkers. The Cry proteins belong to a single family of 50 subgroups. Pests are also killed by lethal septicemia and starvation. An example of a *Bacillus sphaericus*-based product contained a mosquito larvicidal toxin comprising BinA and BinB used for mosquito control.

6.4.7 *Arbuscular Mycorrhizal Fungi (AMF)–Biopesticides*

The composition of AMF varies, and its presence decreases depending on the soil type and crop, as well as the application of fertilisers and tillage. Plants have evolved many direct and indirect mechanisms to overcome herbivory; for example, they produce chemicals such as nicotine, gossypol, and many other compounds that can prevent herbivores from feeding on them. It was discovered that AMF colonisation of crop plants is extremely beneficial in providing a good defence in hosts by altering gene expression patterns and changing the nutritional status of crops directly or indirectly. Some examples of the utility of AMF application in plant biocontrol are already reported, but more research needs to be carried out in this area.

6.5 Nano-Biopesticides

Nanoscale materials-based pesticide delivery is being used for enhanced stability and solubility, enabling relatively slower releases that result in effective targeted delivery in pest management. Organic and polymeric NPs in the form of nanospheres or nanocapsules are used as nanocarriers for herbicide distribution. Polymeric-NPs such as Poly (epsilon-caprolactone) have good properties such as biocompatibility and repeatability after atrazine herbicide encapsulation. In another study, chitosan NPs loaded with three triazine herbicides were shown to exhibit reduced environmental impact and low genotoxic effects in *Allium cepa* (Sanzari et al. 2019).

The nanobiopesticides (size ~ 1–100 nm) are capable of preventing pathogen growth by either destroying or repelling them. Biodegradable nanoencapsulation, nanocontainers, and nanocages increase the stability and efficacy of pest control using smaller amounts of nanobiopesticides.

The phytopathogens that cause damage are overcome by the application of nanobiopesticides, primarily the metallic nanoparticles (NPs) of zinc, gold, silver, nickel, and titanium, owing to their inherent antimicrobial properties. They have added advantages over other biopesticides because of their better solubilization capacities and target-oriented delivery with enhanced efficacy. Bacterial, fungal, and plant extracts are used for the synthesis of NPs. It has been shown that silver nanobiopesticides (Ag-NPs) can be synthesised using marine organisms such as *Sargassum muticum*, *Mesocyclopslongisetus*, and *Caulerpa scalpelliformis*. The advantage of using microorganisms for NP preparation is that they can withstand higher metal concentrations than plants, and their rate of production and management is much faster. The microorganisms, being extremely small, have a better ability to penetrate than plants. A number of microorganism-derived NPs are available for pest control.

Bioherbicides have also been converted into nanoformulations of bioherbicides. The efficacy of metabolites of *Photorhabdus luminescence*, an endosymbiotic bacterium, and *Heterorhabditis indica*, entomopathogenic and parasitic nematodes, has been controlled. Similarly, nanofungicides have also been prepared to control various pathogenic fungi, which include *Bipolaris sorokiniana*, *Fusarium* spp., *Alternaria alternata*, and many others, through Ag-NPs and *Magnaporthe grisea* and *B. sorokiniana* using metal NPs. Apart from their better solubility, the nanofungicides are very economical, eco-friendly, and safe to use.

6.6 Biopesticides from Aquatic Plants

Some aquatic plants like duckweed, muskgrass, water hyacinth (*Eichhornia crassipes*), hydrilla (*Hydrilla verticillate*), water lettuce (*Pistia stratiotes*), and filamentous algae (*Lyngbya wollei*) have the potential to prevent the growth, germination, survival, and reproduction of surrounding organisms. Neem extract kills many insects, while *E. crassipes* has the ability to inhibit the growth of *Spodoptera litura*, a lepidopteran pest. Duckweed and water lettuce inhibit *Chenopodium album*. These examples clearly indicate that such plants or weeds and their allelopathic chemicals are capable of inhibiting the pathogens and hence can be substituted for conventional chemical pesticides, as reported.

6.7 Biopesticides and Chemical Pesticides

Biopesticides have several advantages over conventional chemical pesticides, as they are environmentally friendly, target-specific, and not harmful to non-target organisms, and could thus replace synthetic pesticides for pest management. In recent years, the use of biopesticides has gained popularity because they can be used in sustainable agricultural practices. Biopesticides are highly effective in small amounts, decompose quickly without leaving harmful residues, and thus can reduce the use of conventional pesticides as an integral component of IPM programs. However, despite the merits of using biopesticides, their use has not been as widespread as expected, for the following reasons:

- higher cost of production due to screening, developing, and getting regulatory clearance for each new product;
- shorter shelf-life due to the sensitivity of biopesticides to fluctuations in temperature and humidity;
- Limited field efficacy due to climatic and regional variations in temperature, humidity, soil conditions, and others;
- The higher specificity of the biopesticides is not farmer-friendly. They prefer to use multiple agents to control different pathogens and pests in the field. These agents are confusing, costly, and cumbersome, and they are also not available for every pest or pathogen.

6.8 Commercial Aspects of Biopesticides

A majority of microbial biopesticides available on the market are derived from a single bacterium, *Bacillus thuringiensis* (Bt). Biopesticides account for a small portion of the crop protection market, with a global value of about US$ 3 billion, accounting for only 5% of the total market. In the US market, more than 200 products are available, while the European Union market has only 60 analogues. Biopesticide use on a global scale is increasing by almost 10% every year. However, these pesticides are going to contribute significantly to their global market consumption needs, which are expected to increase further in the future by reducing their over-reliance on chemical pesticides. The EU has been assessing biopesticides using the same regulations as synthetic active substances, which necessitates the addition of several new provisions to current legislation, and the development of new guidelines facilitates the registration of prospective biopesticide products. There are fewer active ingredients of biopesticides registered in the EU than in the USA, India, Brazil, or China. It is expected that the use of biopesticides will be on par with synthetics by the early 2050s, but major uncertainties regarding the rates of uptake, especially in areas such as Africa and Southeast Asia, account for most of the flexibility in such projections, as reported elsewhere (Kumar et al. 2021).

6.9 Chapter Summary

The application of bio-fertilisers consisting of bacteria, cyanobacteria, or fungi can improve and restore the soil's fertility and ensure sustainable agricultural production. Using microorganisms and microalgae as biopesticides can suitably reduce the demand for energy and consumption of synthetic fertilisers and restore the efficiency of agroecosystems and wastelands. These organisms, when combined with innovative RNAi technology, can produce secondary metabolites, biofertilizers, bioenergy, and bioprocessed products that would also be useful in pest control. RNAi-based biopesticides have gained enough popularity in recent years as a narrow-spectrum alternative to chemical-based control that provides specific and accurate targeting of pests and pathogens. In this regard, the use of bioinformatics-based dsRNA selection for effective RNAi design, coupled with precise testing, would most likely eliminate the adverse impacts of RNAi-based biopesticides. Considerable research into biopesticides is required in order for them to enter the biopesticide market in the future. Researchers from diverse fields and institutions around the world are needed to participate and collaborate in making positive efforts in this area. Extensive collaborations between businesses and research institutions are required; otherwise, complete replacement of chemical pesticides with biopesticides appears unlikely. Currently, the agricultural sector relies on biopesticides as well as chemical pesticides. However, translating the laboratory results should expedite large-scale industrial production. The increased inflow of biopesticides, however, has considerably reduced the use of chemicals because of stringent regulations. Many substances have been explored as biopesticides, but extensive field trials are still required in order to assess their efficacy for pest control under diverse cropping systems. Farmers and society at large must benefit from the mixed and judicious use of both conventional chemical pesticides and biopesticides, and it is imperative to emphasize the need to conduct more research in the area of biopesticides in the future.

Further Reading

Chaudhary S, Kanwar RK, Sehgal A et al (2017) Progress on *Azadirachta indica* based biopesticides in replacing synthetic toxic pesticides. Front Plant Sci 8. https://doi.org/10.3389/fpls.2017.00610

Dubois C, Arsenault-Labrecque G, Pickford J (2017) Evaluation of a new biopesticide against angular leaf spot in a commercial operation system. Acta Hortic 1156:757–764. https://doi.org/10.17660/ActaHortic.2017.1156.111

El-Abbassi A, Saadaoui N, Kiai H et al (2017) Potential applications of olive mill wastewater as biopesticide for crops protection. Sci Total Environ 576:10–21. https://doi.org/10.1016/j.scitotenv.2016.10.032

Eski A, Demir D, Sezen K et al (2017) New biopesticide from a local *Bacillus thuringiensis* var. *tenebrionis* (Xd3) against alder leaf beetle (Coleoptera: Chrysomelidae). World J Microbiol Biotechnol 33:95. https://doi.org/10.1007/s11274-017-2263-0

Guerra F, Sepúlveda S (2021) Saponin production from *Quillaja* genus species. Sci Agric, An insight into its applications and biology. https://doi.org/10.1590/1678-992X-2019-0305

Isman MB (2020) Bioinsecticides based on plant essential oils: a short overview. Z Naturforsch 75:179–182. https://doi.org/10.1515/znc-2020-0038

Kadoić Balaško M, Bažok R, Mikac KM et al (2020) Pest management challenges and control practices in codling moth: a review. Insects 11. https://doi.org/10.3390/insects11010038

Kirk WW, Schafer RS (2015) Efficacy of new active ingredient formulations and new biopesticides for managing *Fusarium* root rot disease of gladiolus hybrids. Acta Hortic 1105:55–60. https://doi.org/10.17660/ActaHortic.2015.1105.8

Kumar J, Ramlal A, Mallick D et al (2021) An overview of some biopesticides and their importance in plant protection for commercial acceptance. Plants (Basel) 10. https://doi.org/10.3390/plants10061185

Mauchline AL, Hervé MR, Cook SM (2018) Semiochemical-based alternatives to synthetic toxicant insecticides for pollen beetle management. Arthropod-Plant Interact 12:835–847. https://doi.org/10.1007/s11829-017-9569-6

McGuire AV, Northfield TD (2020) Tropical occurrence and agricultural importance of *Beauveria bassiana* and *Metarhizium anisopliae*. Front Sustain Food Syst 4. https://doi.org/10.3389/fsufs.2020.00006

Moreno-Gavíra A, Huertas V, Diánez F et al (2020) *Paecilomyces* and its importance in the biological control of agricultural pests and diseases. Plants (Basel). 9. https://doi.org/10.3390/plants9121746

Mossa ATH, Abdelfattah NAH, Mohafrash SMM (2017) Nanoemulsion of camphor (*Eucalyptus globules*) essential oil, formulation, characterization and insecticidal activity against wheat weevil, *Sitophilus granaries*. Asian J Crop Sci 9:50–62. https://doi.org/10.3923/ajcs.2017.50.62

Narware J, Yadav RN, Keswani C et al (2019) Silver nanoparticle-based biopesticides for phytopathogens: scope and potential in agriculture. In: Koul O (ed) Nano-biopesticides today and future perspectives. Academic Press, Cambridge, pp 303–314. https://doi.org/10.1016/B978-0-12-815829-6.00013-9

Olowe OM, Olawuyi OJ, Sobowale AA et al (2018) Role of arbuscular mycorrhizal fungi as biocontrol agents against *Fusarium verticillioides* causing ear rot of *Zea mays* L. (maize). Curr Plant Biol 15:30–37. https://doi.org/10.1016/j.cpb.2018.11.005

Pavela R, Waffo-Teguo P, Biais B et al (2017) *Vitis vinifera* canes, a source of stilbenoids against *Spodoptera littoralis* larvae. J Pest Sci 90:961–970. https://doi.org/10.1007/s10340-017-0836-1

Ranglová K, Lakatos GE, CâmaraManoel JA et al (2021) Growth, biostimulant and biopesticide activity of the MACC-1 *Chlorella* strain cultivated outdoors in inorganic medium and wastewater. Algal Res 53:102136. https://doi.org/10.1016/j.algal.2020.102136

Rodrigues D, Costa-Pinto AR, Sousa S et al (2019) *Sargassum muticum* and *Osmundea pinnatifida* enzymatic extracts: chemical, structural, and cytotoxic characterization. Mar Drugs 17. https://doi.org/10.3390/md17040209

Saberi F, Marzban R, Ardjmand M et al (2020) Optimization of culture media to enhance the ability of local Bacillus thuringiensis var. tenebrionis. J Saudi Soc Agric Sci 19:468–475. https://doi.org/10.1016/j.jssas.2020.08.004

Sanzari I, Leone A, Ambrosone A (2019) Nanotechnology in plant science: to make a long story short. Front Bioeng Biotechnol 7. https://doi.org/10.3389/fbioe.2019.00120

Chapter 7
Nanotechnology and Crop Management

7.1 Introduction

Crop management helps improve crop productivity, leading to higher yields with improved quality. Crop management is a set of agricultural practices performed to improve the growth, development, and yield of crops. It comprises steps like seed-bed preparation, sowing of seeds, and crop maintenance; and ends with crop harvest, storage, and marketing.

Fertilization is an important component of crop management. It is better to test the soil for available plant nutrients before adding fertilizers to any crop. The addition of appropriate fertilizers determined from the soil and/or plant analysis can ensure the planted crop's nutritional requirements. The amount of fertilizer, type, form, timing, and method of application are all determined by a variety of factors, such as crop and fertilizer types, soil conditions, and weather conditions. The previous crop and past manure applications also influence crop nutrient needs.

Pest management is another important aspect of crop management. Pesticides are used for controlling pests in most crops, mainly if used correctly based upon specific pest species. Additionally, integrated pest management practices can provide growers with an economical option that is safer and often more beneficial to humans and natural resources.

7.2 Plant Healthcare: Different Aspects

There is a critical demand for more sustainable agricultural production by increasing crop yields to meet the demand of a rapidly growing population. The UN estimated that by 2050, the global population is expected to reach 9.8 billion people. However, farmers are facing many problems, such as extreme temperatures, soil

A. Nazeer et al., *Targeted Delivery of Nanopesticides and Nanofertilizers in Sustainable Agricultural Farming*, Nanotechnology in the Life Sciences,
https://doi.org/10.1007/978-3-031-41333-9_7

degradation, and drought, that are expected to worsen as the climate changes. Increased sustainable agricultural practices are therefore becoming essential to introduce to ensure high yields that utilize minimal inputs and are least destructive to the land.

Plant health monitoring is one such method to increase yields and decrease environmental impact by deploying low-cost and in-field methods to constantly monitor water level, soil quality, and the presence of pathogens and pests. Expensive agrichemicals and water can be used in a directed manner for optimal plant growth. Pathogen detection would allow for immediate corrective action to prevent disease from spreading. There are many agricultural practices and technologies currently employed by farmers to maximize yields, such as crop rotation to improve soil health, the use of genetically modified seeds, or monitoring plants for the presence of pathogens and pests by planting non-native plants or sentinel plants. There are many diagnostic technologies employed to detect disease. However, current laboratory-based techniques for plant diagnostics are not adequate for point-of-use plant monitoring. There are several point-of-use technologies that have been developed, such as lateral flow devices or portable devices for in-field use. However, these types of devices require harvesting and processing plant tissue, which is not conducive to continuous monitoring.

Nanotechnology in plants is an emerging field from the past decade that has the potential to create more productive agricultural systems. The use of nanotechnology has been extensively studied for applications in human health, medicine, pharmaceuticals, and wearable devices. Even implantable sensors for continuous monitoring in humans are possible. Nanotechnology has the potential to improve agriculture in several ways, including the formulation of nanofertilizers and agrichemicals, novel delivery mechanisms for agrichemicals, nanosensors for disease detection, nanodevices for genetic modification, and postharvest crop management. Emerging technologies for *in vivo* plant sensors are upcoming for monitoring plant health (Roper et al. 2021).

7.3 Genetic Engineering Approach

7.3.1 Synthetic Biology

One class of *in vivo* plant sensors, called phytosensors, was developed using synthetic biology. Phytosensors are plants that report plant pathogens, toxins, or nutrients. Plants have an innate, inducible defence mechanism to protect against pathogens, toxins, and nutrient deficiencies. Phytosensors are created by fusing reporter genes, such as fluorescent proteins, to synthetic inducible plant defence promoters. By fusing reporter genes to plant stress promoters, plants sense pathogens at a molecular level and quickly have a visible-to-the-naked-eye readout. This allows for rapid detection, as it is often several days or weeks from the point of infection to the presentation of visible symptoms. Since plants naturally sense biotic and abiotic changes and alter biochemical and gene expression patterns, phytosensors hold promise as a modular and easily modified biosensor. This type of sensor is feasible for on-the-ground, in-field detection or could be used on a larger scale to monitor fields via satellite images with image detection software. There are several proof-of-concept studies. The elements from the promoter regions of pathogen-inducible genes and genes responsive to plant defence signal molecules such as salicylic acid, jasmonic acid, and ethylene were used on *Arabidopsis* and tobacco as their model hosts and transformed them with the pathogen-inducible synthetic promoters fused with the reporter gene, GUS. Phytohormone and plant elicitor treatments showed that the expression of GUS was increased compared to that under control. Transformed tobacco plants had increased expression of GUS when infected with Alfalfa mosaic virus but not Tobacco mosaic virus, demonstrating that different promoters could be used to detect different targets. In another study, four pathogen-inducible promoter elements were fused to orange fluorescent protein in *Arabidopsis* and tobacco, followed by testing the robustness and predictability of the transgene by monitoring the transgenic tobacco throughout two field seasons. They found that 3 out of 4 transgenic lines maintained the expected fluorescence signal. One line, in particular, was specifically induced by bacterial phytopathogens and showed an increase in fluorescence only 48 h after infection, much sooner than visible symptoms. These studies demonstrate the feasibility of phytosensors in live plants and in field settings. There are many innate plant responses that could be used in the design of phytosensors, though the degree of specificity and sensitivity would vary greatly among each promoter and element and would require widespread studies reported recently (Roper et al. 2021).

7.3.2 *Imaging and Spectroscopic Approaches*

Imaging and spectroscopy are two other methods of rapid diagnostics. Molecular methods that use spectroscopy, like real-time PCR and ELISA, are common methods for plant disease diagnostics but are highly invasive. Imaging includes techniques such as thermography, RGB imaging, fluorescent imaging, and hyperspectral imaging.

7.3.2.1 Imaging

Thermography imaging detects heat emitted by objects, and it is often used to survey large stretches of land at once. Changes in plant temperature can be attributed to several factors, including pathogen responses, such as closing stomata, or abiotic stress. While this method is ideal for monitoring large fields and is non-invasive, it is an indirect and non-specific detection method.

RGB imaging utilizes digital cameras to measure any changes in transmittance. Simple digital images and videos have been used for monitoring a diverse set of plants in a field. It can be used for single plants, such as with a smartphone sensor, or used with drones to monitor large fields. Notably, machine learning algorithms are being deployed to detect patterns that indicate disease. Since RGB imaging relates changes in colour to changes in plant health, it is an indirect method and cannot always provide specific insight into factors affecting the plant.

Fluorescent imaging is like RGB imaging; however, it often includes a laser in addition to a camera for fluorescent excitation. The most common use of fluorescent imaging is chlorophyll fluorescence imaging, where the fluorescence of a leaf or plant is compared to that of surrounding plants or to a baseline value. Chlorophyll naturally fluoresces when excited by certain lights. Several studies have used this phenomenon by relating fluorescence to the activity of photosynthesis. A chlorophyll fluorescence model was developed to determine the photosynthetic activity of leaves collected from areas with different ambient air pollution and different agrichemical treatments. These different factors altered the chlorophyll fluorescence, indicating some interruption in photosynthetic activity. This study also outlined several possibilities for portable in-field devices. Since chlorophyll is fluorescent under intense sunlight, a simple fluorimeter can be used to take measurements in the field. Even though this method is non-invasive, non-destructive, and easily adaptable to in-field use, it is non-specific and cannot diagnose specific abiotic or biotic stressors. Leaf fluorescence fluctuates often and in response to multiple biotic and abiotic factors.

Hyperspectral imaging is a technique that analyses light across the electromagnetic spectrum to evaluate changes that are not always visible in RGB images. Though it can detect more nuanced changes than visual or fluorescence images, it can only be used to detect general changes in plant surfaces. With further studies, hyperspectral patterns can be attributed to specific conditions. For example, hyperspectral features of yellow rust disease were analyzed statistically to differentiate yellow rust from nutritional deficiencies.

Subsequent studies involved the use of polydiacetylene (PDA) polymer and DNA-functionalized single-walled carbon nanotubes (SWCNTs) incorporated into leaves before imaging. Both techniques were solely carried out in a lab setting, though both show promise of potential in-field applications that incorporate materials directly into live plant leaves for diagnostics. To measure the amount of water output from individual stomata, a PDA-based brush-on sensor method was reported with a hydro chromic PDA system. Diacetylene monomers were brushed on the abaxial side of the leaf and photopolymerized, followed by a fluorescence microscopic study to detect the change in moisture, as the polymer undergoes a blue-to-red transition in response to changes in moisture coming from individual stomata. With fluorescence microscopy, open stomata can be detected to see possible environmental effects (temperature, wind, or humidity) on stomata activity. This is a small-scale, lab-based application that has the potential to be used for in-field diagnostic methods. A hydrogen peroxide sensor was reported based on functionalized SWCNTs and near-infrared fluorescent imaging. Hydrogen peroxide is generated in response to plant stresses. In this study, the effects of UV-B, high light, wounding, and pathogen-related stresses were tested, in addition to the direct application of hydrogen peroxide. The SWCNTs were functionalized with the aptamer sequence that binds to hemin, which catalyzes hydrogen peroxide to produce hydroxyl radicals. The reactive hydroxyl radicals then quenched the SWCNTs' fluorescence in the near-infrared range. In conditions of direct hydrogen peroxide application and in stressful conditions, fluorescent emissions were reduced. This nanosensor was thus able to provide early signs of stress and could be optimized for precision agricultural practices and monitoring plant health. SWCNTs can be functionalized using varying methods for the detection of a wide variety of analytes (Roper et al. 2021).

7.3.2.2 Spectroscopy

Raman spectroscopy has been found useful in detecting the vibrational frequencies of molecules, which can then be used to determine the chemical footprint of a structure and thus identify the molecules. Simply, a sample is illuminated with a monochromatic laser that interacts with the sample, and the resulting shift in energy gives insight into the presence of molecules within the sample. Raman spectroscopy is non-destructive and biochemically safe for the detection of molecules in highly complex samples. Subsequently, a portable Raman spectroscopy instrument was developed for use with *Coleus* Lime as their model organism. Two photosynthetic pigments, anthocyanins and carotenoids, were the target molecules for the Raman study. Carotenoids are a first line of defence against ROS, and anthocyanins block harmful irradiation. Both increase biosynthesis in response to several environmental factors. Four methods of abiotic stress were applied: light irradiation, cold, drought, and saline stress. Using both a Raman microscope and the portable Raman instrument, the relative concentration of carotenoids and anthocyanins was assessed, indicating abiotic stress, and was determined 2 days after the application of light, cold, drought, and saline stress. The concentration of carotenoids and anthocyanins indicated the presence of stress in the plant before physical symptoms arose. Both results were confirmed with chemical analytical extractions. The changes to these pigments over time showed that Raman spectroscopy was a method to accurately measure these molecules and indicated a functional relationship between the molecules and their response to excessive ROS during abiotic stress. The portable Raman instrument, however, was unable to detect anthocyanins. Further optimization could expand the capabilities. A portable Raman leaf clip sensor was reported to distinguish between nitrogen-rich and nitrogen-deficient plants. Raman spectroscopy was also shown to detect pathogens and pests that live within host seeds and the presence of chemical pesticides.

X-ray fluorescence (XRF) spectrometry is a non-destructive method used to determine the chemical composition of many sample types. In XRF, an X-ray beam interacts with the sample, and the fluorescent X-rays produced can be used to identify the elements in the sample. XRF was used along with an infrared gas analyzer to elucidate the uptake kinetics of aqueous Zn and Mn in soybean leaves and stems for 48 h. The elemental distribution changes in plants were measured to see the effect of localized X-ray exposure on live plant tissue. Typical XRF did not cause visible damage, dehydration, or elemental redistribution in live plants, though the long-term effects of low-dose X-ray exposure have not yet been studied.

Mass spectrometry is used to determine the mass-to-charge ratio of ions, and there are several different types depending on the sample to be analyzed. Ambient ion mass spectrometry allows for mass analysis without typical sample manipulation, such as a high vacuum environment. Low-temperature plasma (LTP) can be used to ionize samples in ambient air. LTP is a gentle method of ionizing to be used to analyze nicotine biosynthesis in live tobacco plants (Roper et al. 2021).

7.3.2.3 Combination Methods

There is a recent trend toward combining two or more imaging or spectroscopy methods for more accurate diagnostics and more sensitive detection.

A method reported allowed for *in vivo* monitoring of genomic targets by integrating plasmonic nanoprobes and three complementary techniques to image and sense the probes: surface-enhanced Raman scattering (SERS), XRF, and plasmonic-enhanced two-photon luminescence (TPL). Using plasmonic-active silver-coated gold nanostars functionalized with double-stranded DNA, it was noted to change conformation in the presence of a specific biotarget. These probes were used to detect miR156, a miRNA in *Arabidopsis*, but they could be used to sense a wide variety of biotargets as well. The technique was validated in *Arabidopsis* using SERS tags to verify agreement among imaging modalities. Nanoprobes were used to detect miR156. Raman imaging only detects the probe when it binds to its target. TPL and XRF detect the probe regardless of target interaction. The XRF signal was used to normalize the signal from Raman spectroscopy, allowing for quantification, an important aspect of biosensing. This method was not only used to track changes in each target over time, but it was also used for plant pathogen diagnostics. In other studies, thermal imaging and fluorescence imaging were complementary to each other in monitoring for plant stress (Roper et al. 2021).

7.3.2.4 Electrical-Based Approaches

Several studies have been conducted that use electrical components for *in vivo* plant monitoring. While this requires external equipment, the use of nanotechnology allows for devices that can be integrated into plants.

Microneedle Electrodes

Measurement of salinity has been considered an important factor in plant health and crop yield, for which a real-time monitoring system was developed to detect salinity in a non-destructive manner through electrical conductivity inside the stems of tomato plants. A self-contained unit was developed, including a microneedle electrode and electrode pad that were inserted into the stem of a tomato plant kept under greenhouse conditions and in field conditions. In field conditions, there was a decrease in signal noise and a decrease in electrical conductivity measurements. A similar methodology, employing a thermal microneedle probe, was used to measure xylem sap movement in tomato stems. Taking maize as a model, a system was developed using a similar microneedle leaf sensor. However, the device could measure canopy temperature for water stress assessments. It can be calibrated for any plant, soil type, and relative humidity. It is powered by solar energy and emits data wirelessly through an antenna (Roper et al. 2021).

Organic Electrochemical Transistor Sensors

An organic electrochemical transistor sensor (OECT) was developed to explore biosensing. Simply, a conductive polymer film or channel was placed in direct contact with an electrolyte and electrodes. There are source and drain electrodes connected to the channel, and a gate electrode that establishes an electrical connection to the electrolyte. The conductive polymer poly(3,4-ethylenedioxythiophene) (PEDOT) doped with various side groups is used to make a common OECT sensor.

An OECT sensor was reported for continuous monitoring of plant health based on changes of solutes in sap used in tomato as a model organism, as commercially grown tomato requires optimization of conditions throughout its cropping cycle and yield and quality is largely variable. Here, OECT sensors are integrated into plant stems using cotton fibers. These sensors were highly biocompatible and commonly integrated into textiles to detect sweat. Commercial cotton fiber was functionalized by soaking in the conductive polymer and drying in the oven. Functionalized cotton was inserted into the tomato stem and cut so it protruded from each end of the stem. A thin metal wire was attached to either end of the cotton thread, and a third thin wire was introduced as the gate electrode. Time constant and resistance (based on voltage across the sensor) were measured to deduce the physiological state of the plant. Although this is an indirect measurement, it could be used to continuously monitor over time. Recently, use of this kind of sensor was reported for drought detection in tomato plants. Using a bioristor sensor, drought stress was detected only 30 h from withholding of water. OECT sensors were used for measuring the glucose and sucrose levels in the xylem sap of aspen trees (Roper et al. 2021).

Handheld Plant Health Sensor

Atoptix is developing a handheld, smartphone-based plant health sensor that can provide real-time quantitative measurements of crop nutrient levels and water stresses, as well as early flagging of diseased plants that are not yet visually distinguishable. The plant health sensor is built on a patented miniature, low-cost, and yet high-resolution spectroscopy platform. The spectrometer technology was originally developed by Atoptix co-founders at Penn State. The sensor is unique as it essentially uses the plant as the gauge stick for determining agro-chemical application rates, in contrast to a nearby soil measurement or a subjective visual inspection approach.

The handheld plant health sensor is designed for use in the precision agriculture market, capturing optical measurements from the leaf of a plant that can be used to diagnose the plant's overall health and provide recommendations for optimizing care. In discerning whether a plant suffers from nutrient deficiency, water stress, or disease pressure, the sensor provides a means for farms to better manage and substantially save agrochemical and irrigation resources. Additionally, the ability to time stamp and geo-stamp spectral data through integration with smartphones enables the development of an extraordinary spectral database at a scale never before possible. Such a capability brings the power of big data analysis to a myriad of applications in the agricultural and environmental markets.

Atoptix's current focus is to bring a product to market for early disease diagnosis in high-value crops, particularly for diseases without a cure such as the deadly citrus greening disease. Atoptix has developed functional prototypes of the plant health sensor and is currently performing field testing with the devices to demonstrate the early diagnostic capability.

The core spectrometer technology was patented by PennState (US Patent 8,861,086 and foreign patents pending).

7.4 Chapter Summary

There are a diverse set of needs for better plant diagnostic technologies. The best technology for a given farmer will depend on the size of the land they are farming, the specific needs of their crops, and the natural, social, and economic environment they are in. Developing an array of sensors and innovative technologies is important for meeting the agricultural demands of a larger population. Current technology for measuring plant health or diagnosing disease is expensive, invasive, and often requires sending samples to central facilities for processing. Nanotechnology and advanced spectroscopy techniques are emerging diagnostic technologies for plant disease and detecting plant distress, all with the common goal of increasing yield in a sustainable way. The current challenges of these technologies include field implementation. Many of these studies are proof-of-concept demonstrations and would require further studies to determine their efficacy in the field. Factors important to consider for a successful *in vivo* sensor include, but are not limited to, accuracy, specificity, sensitivity, durability, cost, ease of use, and environmental impacts. These sensors could certainly allow for precision agriculture, where expensive resources are used in a directed manner and crop yield is maximized. Moreover, making these technologies affordable and accessible to large-scale and small-scale farmers alike is vital, as both are important in increasing agricultural production.

Further Reading

PennState Materials Research Institute. https://www.mri.psu.edu/mri/innovations/innovations-4. Accessed 15 Dec 2022

Roper JM, Garcia JF, Tsutsui H (2021) Emerging technologies for monitoring plant health *in vivo*. ACS Omega 6:5101–5107. https://doi.org/10.1021/acsomega.0c05850

Chapter 8
Targeted Delivery of Nanopesticides

8.1 Introduction

Nanobiotechnology with extensive applications in ecological sustainability has been emerging as an integrated approach for supporting the agro-economy and ecosystem service providers. The synthesis of nanoparticulate material species using living organisms like bacteria, fungi, and plant extracts and their metabolites has already been demonstrated and reported. Biogenic NPs being green alternative to the existing chemical methods of nanoparticulate syntheses that are toxic and energy-intensive with hazardous byproducts that damage the environment. Biosynthesized nanoparticles, on the other hand, are biocompatible, highly reproducible, have lower polydispersity, and are simple to scale up for commercial use. The biological NP-syntheses are comparable to other existing physical or chemical methods. Microorganisms, with their inherent properties to remediate and reduce heavy metals via numerous reductase enzymes, have the potential to act as nano-bio-factories for synthesizing metallic bio-NPs. Over the years, interest has moved from inorganic to biogenic NPs in the form of nanocomposites. Nano-agrochemicals employing biogenic delivery frameworks compete well in terms of environmental sustainability but often do not compete economically with other agrochemicals. Investigations are therefore becoming necessary to determine whether these nanoproducts could compete with the existing chemical formulations in terms of performance and cost. The biosynthesis of NPs can be broadly classified into two categories: intracellular and extracellular.

Intracellular syntheses occur within the biomass (cells) of plants, fungi, and bacteria, while extracellular ones occur outside the organism, which is in general aided by several biomolecules and extracellular metabolites. The challenges of intracellular synthesis, including the high extraction cost and complex downstream processing, could be overcome by extracellular syntheses, thereby finally improving the efficiency of downstream applicability. In the face of bioremediation,

A. Nazeer et al., *Targeted Delivery of Nanopesticides and Nanofertilizers in Sustainable Agricultural Farming*, Nanotechnology in the Life Sciences,
https://doi.org/10.1007/978-3-031-41333-9_8

intracellular synthesis is preferred, as it is easier to remove the microbes containing the contaminant. For all NP synthesis methodologies, a good monodispersive feature can be achieved by controlling the parameters like metal salt concentration, incubation time, pH, and temperature. The biosynthesis of NPs helps in providing characteristic natural capping to NPs and thus provides additional stability, which can enhance the efficacy of biological NPs. Several organisms known to synthesize NPs by both extracellular and intracellular strategies have been selected for intensified NP-productivity using simple upscaling, a higher surface area of reactivity, and the amount of proteins produced and excreted by organisms for effective downstream handling (Ahmad and Ahmad 2021).

In a study conducted in an inert growth medium, the application of hydroxyapatite nanoparticles stabilized by carboxymethylcellulose increased the growth rate and seed yield of soybeans by 33% and 18%, respectively. However, changes in grain production due to the application of hydroxyapatite-NPs were not reported in the study, and no significant statistical differences were observed while comparing the application of traditional phosphorus fertilizer and hydroxyapatite-NPs. It must also be considered that the results expected from real soil conditions were different from those observed under controlled laboratory conditions. Synthesis of nanofertilizer form of nitrogen was reported recently synthesized wherein hydroxyapatite-NPs were covered with urea, leading to a slow release of nitrogen to the plants. Additionally, research in rice plants indicated that urea nanohybrids delivered nitrogen multiple times over a period of time, much slower than synthetic urea, with an enhanced grain yield at just half the rate utilized with synthetic urea. In another study, phosphorus-NPs applied as a seed treatment in a suspension of water and phosphorite increased the growth rates in nine test plants of wheat, rye, pea, barley, corn, buckwheat, tomato, radish, and cucumber by about twofold. Similarly, the high solubility of phosphate due to the use of ammonium zeolites increased the uptake of phosphorus in plants. The fungal-mediated biosynthesis of phosphorus-NPs was carried out employing *Aspergillus tubingensis* using tri-calcium phosphate as a precursor salt (Camara et al. 2019).

A slow-release nano-potassium fertilizer formulation was developed and found to reduce potassium losses in the soil while supporting the long-term sustained supply of potassium to plants. Likewise, the foliar application of nano-potassium fertilizer improved the biomass, growth, and yield of parchment pumpkin. In another study, zinc NPs improved the zinc content in the grains and enhanced the growth and yield of maize growing in Zn-deficient soil. However, the study was performed in sandy clay loam with a low organic matter content, and different results could be observed if the organic matter content was increased even in the same soil. In another study, the use of ZnO-NPs increased the lycopene content by 113% in tomato. Similarly, the use of Fe-NPs in various crops improved enzyme function and overall agronomic traits, including Fe biofortification, nutritional quality, biomass, and N and P metabolism. In another study, the use of ZnO-NPs in cluster beans increased the P uptake by 11%, as the Zn helps in mobilizing P for plant use. The application of ZnO-NPs increased the cotton biomass and yield and increased the activity of antioxidant enzymes and other regulatory and functional proteins in the soil. Similarly, P mobilization occurred with the use of magnetite NPs (Fe_3O_4) in lettuce. Thus, along with serving the primary role of a nanonutrient, NPs have multi-dynamic and systemic effects, with an active role in mobilizing and increasing the availability of other nutrients and improving the overall nutritional quality of farm produce.

The use of nanotechnology in plant sciences is gaining interest, particularly for the use of NPs as vehicles for biomolecules of agronomic importance or agrochemicals in plants. Nanoparticulate encapsulation of nutrients in a nano-thin protective film or nanoemulsion ensures a stronghold of nutrients on the plant surface due to the higher surface tension of the nanocoating and has great potential for improving crop efficiency. Nanozeolites and nanoclays are typical examples of products used as soil improvement products to aid in the efficient release and retention of water and nutrients. These zeolite-NPs have characteristically well-defined pore networks that ensure the slow release of the agrochemical or nutrient in question. Similarly, nanoscale materials in the form of nanowires (including microbial nanowires) and nanofibers help in the development of nanosensors and related diagnostic devices for the detection of pesticides and fertilizers. Hence, nanobiotechnology is a key enabling technology that could revolutionize modern agriculture to a large extent.

Various factors come into play during the uptake and translocation of NPs inside the plant system, from physiological plant factors—such as age, the plant species itself, and its biotransformation pathways—to the biophysicochemical factors of NPs, which define the functionalization of NP when presented to plant cells and have a cumulative effect on their fate. The most commonly used method for providing nutrient supplements is the soil application of organic and chemical fertilizers. Soil is a dynamic and heterogeneous mixture of several biotic and abiotic factors (living or dead) that poses challenges with respect to various soil properties such as texture and pH, which govern the fate and longevity of fertilizer in the soil. Ion exchange capacity also plays a vital role in nutrient mobilization in the soil-plant system.

Another alternate method available uses liquid fertilizers sprayed directly onto the aerial plant parts, mainly leaves. This minimizes the loss of nutrients and makes them readily available for plant use, circumventing the soil issues that make them less available to the plants. Despite its clear advantages, this method requires rigorous optimization by considering the role played by stomata and epidermal cells in nutrient uptake, bearing in mind their diurnal physiological responses. Another critical aspect to be taken into consideration is the size of the nutrient formulation, which should not interfere with or hinder normal stomatal function by blocking the pores in the event that the nutrient supplied is larger than the size of the stomatal pore or if it is provided in an excess concentration.

Similarly, nanoagrochemicals are generally delivered to plants via three methods: seed treatment, soil application, and foliar application. In general, the possible modes of entry of NPs into the aerial parts of the plants include passive uptake through plant openings with specific size exclusion via stomata, bark, and hydathodes. However, other physiological and anatomical perspectives should be considered to better understand how NPs and plants interact.

Other viable routes for NP-uptake include wound and injury on plant surfaces. Lateral root junctions in the rhizodermis, especially near the root tip, provide easy access to the NPs at the root level. In general, the presence of microbes (symbiotic or parasitic), organic matter, and exudates, among others, in the soil complicates the dynamics of NP-uptake relative to that in aboveground plant parts. In case NPs are applied to soil rather than using foliar spray, the high exposure of these NPs may influence soil microbial communities and agglomeration of NPs due to various soil physicochemical properties that could confine NP uptake by plants. It is evident from the published literature that NP-delivery by foliar application is more advantageous for nanonutrient uptake. Furthermore, laboratory scale tests have revealed that aerosol spray helps to produce monodisperse particles that do not agglomerate when applied through foliar application (Camara et al. 2019).

Excess amounts of minerals present in the soil limit plant growth and regulation processes—metals accumulate in the soil, which in turn affect water availability and plant growth. Roots are an integral point of contact between plants and soil, facilitating the uptake of nutrients, minerals, water, and other biologically relevant substances. Root development depends on the aeration, nutrient availability, pH, and texture of the soil, all of which affect nutrient availability to the plants. The uptake and transfer of nutrients from the soil through roots to shoots mainly occurs by diffusion and bulk mass flow.

Diffusion aids the movement of nutrients along the concentration gradient, whereas bulk or mass flow involves the movement of solutes and water together (through xylem and phloem at the whole-plant level) by the pressure gradient and is dependent on the transpiration rate and nutrient availability in soils. The NP-uptake is difficult to predict as it depends on various factors like size, net charge, surface coating, and surface functionalization, and stability besides the application route (aerial, root, and seed) and interactions with the environmental components (soil surface, water accessibility, and microbiota), cell wall rigidity, and the physiology, anatomy, and biochemistry of the individual plant species.

After entering the plant's external defensive layers, NPs have two routes of mobilization: the apoplastic and symplastic pathways. Apoplastic transport is found advancing radially, which tends to move NPs toward the central cylinder of the root and vascular tissues leading to an upward trend toward aerial parts. Apoplastic translocation is vital for applications that require systemic delivery of NPs. The Casparian strip prevents the radial movement of NPs in the root endodermis, which can be overcome by switching from the apoplastic to the symplastic pathways. The symplastic pathway is a more regulated and organized for facilitating the movement of NPs in the plant system. Further, NPs can enter cells by different processes, including phagocytosis, pinocytosis, or endocytosis, and can accumulate in the cytoplasm, vacuoles, or lysosomes. The receptor-bound endocytosis is a significant interfacial phenomenon between NP-bound protein epitopes/ligands and integral receptors bound with the cell membrane, which determine NP contact with cells and engulfment at particular (generally specific) adhesion sites. In addition, the direct cell passage of NPs that does not hamper cell integrity can be achieved by using amphipathic CPPs (cell-penetrating peptides). The cationic groups (CPPs) communicate with membrane-bound anionic groups on one side, leaving their hydrophobic surface to reach the hydrophobic insides of the film easily, subsequently overcoming the hydrophobic barrier presented by cell membranes. Cationic strength over dosage can hamper the integrity of the cell layer, consequently prompting cytotoxicity and eventually cell lysis (Camara et al. 2019).

Once the NP is in the cytoplasm, cell-to-cell movement occurs with the help of plasmodesmata. Studies have revealed that metal-NPs can infiltrate seeds and translocate into the seedlings with no significant effect on germination rate or viability, suggesting the effective use of functional NPs for stimulating plant growth using seed priming.

One of the current challenges facing global agriculture is the need to reduce the intensive use of agrochemicals such as herbicides, fungicides, and insecticides. These substances are well-known for their adverse environmental and social impacts, as well as their ability to induce resistance in target organisms. Large amounts of pesticides are moreover lost during application due to volatilization, degradation, and photolysis, with less than 0.1% of applied pesticides effectively acting against the target organisms. In this context, it is expected that using nanotechnology would offer solutions for mitigating the harmful effects of pesticides on the environment and human health, since in these systems it might be possible to deploy the controlled release of active compounds, resulting in enhanced efficiency and safety while reducing the overall quantities required in field applications. Controlled release systems would thus be leveraging the emerging technologies that have attracted global commercial and scientific interest recently. Such systems are used in areas including medicine, cosmetics, engineering, food, and agriculture. Considering the importance of using them in the agricultural and environmental sectors, a 10-year survey (2009–2019) showed that only 6% of research concerning nanoformulations was related to the agroindustry field, of which 77% was related to controlled release nanoformulations while only 23% involved stimuli-responsive materials. Out of various stimuli-responsive materials, those in the area of pH-based release were the most widely studied (37%), followed by photo and thermal stimuli (27% and 17%, respectively). Enzymes, redox, and other types of stimuli put together accounted for 20% of the studies. The commercial trends of micro- and nano-formulations have been investigated by analysing patent data (Camara et al. 2019).

The advantages of controlled release-based targeted delivery of agrochemicals include reductions of phytotoxicity, leaching, volatilization, drift, and soil degradation, as well as safety during applications. However, the release of active agents from loaded particles mainly occurs due to passive diffusion, capsule erosion, or osmotic pressure, resulting in poor control of pesticide releases. Hence, stimuli-responsive systems were studied for improving the controlled release properties and promoting site-specific and smart release of pesticides in response to biotic or abiotic stimuli.

Biotic (i.e., plant pathogens, insects, and weeds) and abiotic (i.e., temperature, drought, flooding, and salinity) stressors are found to cause a reduction in global crop productivity of about half. Therefore, exploring newer responsive carriers that respond to environmental stimuli such as pH, temperature, redox conditions, light, and enzymes could be considered a better strategy for improving the resistance of crops to the stresses caused by these environmental factors. The formulations using stimuli-sensitive carriers is thus found effective in improving the controlled release system, where the active responses of macromolecules to small signals or modifications in the surrounding environment, would be affecting more changes in their physicochemical properties favouring the release of loaded substances. When used in agrochemical formulations, these sensitive carriers would thus enable the release of active compounds in response to abiotic or biotic stimuli. Stimuli-controlled delivery systems can be prepared using natural polymers like alginate, cyclodextrin, starch, chitosan, carboxymethylcellulose, ethylcellulose, and silica due to their wide availability and biodegradability. Carriers based on pillararenes and curcubit[n]urils are currently being examined in the context of pesticide and herbicide delivery due to their better responses to the environmental stimuli. Thus, many physicochemical modifications could be incorporated in the polymer structure, including changes in physical state, shape, conductivity, solubility, hydrophilic/lipophilic balance, and sol-gel transition. These reversible changes incorporated in the release systems of different stimuli-responsive NPs will be of significant importance. The use of stimuli-sensitive carriers in nanoformulations can be effectively used in the release of different compounds at specified times, thereby increasing bio-efficacy while reducing side effects, the dosage required, and the number of applications. The use of stimuli-responsive-NPs offers a promising technology that is still in an early stage of development and clearly needs further explorations to extend its use in agro-industry. This chapter is accordingly devoted to discussing the current state of development of stimuli-responsive carrier systems, where the mechanisms of release of active ingredients are activated by different triggers in response to biotic and abiotic factors, as well as the perspectives and commercial challenges for the use of smart delivery systems in the agricultural sector (Camara et al. 2019).

8.2 Tracking of NPs

NPs should be thoroughly characterized for their applications and biosafety. Physicochemical characterization is essential for understanding the material's properties, qualities, and functionalities, such as solubility, dispersibility, and stability. The recent interest in metal NPs has made it important to develop methods for studying their uptake, transformation, and environmental fate, hence raising the need to design tools for the real-time tracking and visualization of engineered nanoparticles (ENPs) in the soil-plant system and advanced microscopic and tracer techniques.

Some studies have used micro-X-ray fluorescence and absorption spectroscopy (i.e., μ-XRF and μ-XAS) to study the translocation and accumulation of NPs in the plants. The characterization and particle size distribution of NPs in leaves can be investigated using single-particle inductively coupled plasma mass spectrometry (sp-ICP-MS), and the spatial distribution of metals in leaf tissues can be analysed using X-ray fluorescence microscopy (XFM). Metal speciation can be assessed with X-ray absorption near-edge spectroscopy (XANES) without damaging the samples. In situ analyses of metal speciation in plant tissues have been carried out using synchrotron-based X-ray absorption spectroscopy, which can reveal the underlying mechanisms around nutrient bioavailability, mobility, and behaviour. Translocation of TiO_2-NPs in cucumber was, for instance, analysed using synchrotron μ-XRF and μ-XANES. For tracer techniques, NPs are tagged with suitable probes for analysis using positron emission tomography (PET) or single-photon emission computed tomography (SPECT) to visualize inside plant tissues. The labelling of metal NPs with radioactive isotopes is another valuable tool for the sensitive and specific localization of NPs in a system. Labelled NPs have a similar surface and chemical composition as unlabelled NPs and thus reflect the actual behaviour of NPs, which is lost in other fluorescence labelling techniques due to the surface modifications required therein.

The intrinsic radiolabelling has more advantages than the extrinsic radiolabelling because the exogenous radiotracer in exogenous labelling is prone to breaking away from the metal core of the NP, such that it does not represent the actual behaviour of the metal-NP and is a misfit for *in vivo* studies. *In situ* tracking and imaging can be achieved with intrinsic radioactive isotope-labelled NPs using nuclear imaging techniques (PET or SPECT). These advanced tracer techniques are expected to help in clarifying phenomena like adsorption, translocation, transformation, and assimilation of NPs in the soil-plant system in terms of cellular uptake of NPs and parameters controlling nanocellular interactions.

8.3 Smart Target Delivery

Based on environmental triggers and biological demands, it has been found possible to improve the commercial production of physiologically and agronomically important metal core NPs for fertilizer formulations to curb nutrient losses and increase plant use efficiency, and thus qualify as "smart delivery systems." As nanofertilizers, NPs can be applied to plants or soil to enhance and improve fertiliser uptake, resulting in improved performance. While nanofertilizers are promising for agribusiness, the use of nanotechnology for nutrient supply is sparsely developed. Multifunctional characteristics of the constituent define a smart delivery system with a highly controlled, specific, remotely coordinated, and targeted approach that avoids biological barriers. Likewise, nanofertilizers can be discharged at much slower rates, assisting soil fertility and nutritional balance by reducing run-off into groundwater and minimizing the toxicity caused by over-application. Zeolites, having nano-porous structures, exhibit high selectivity towards plant nutrients and a high specific surface area, which might be released in a slow and controlled manner for uptake by the plant as per the need, thereby ensuring minimized nutrient loss and ecological risk and improved viability. Nanofertilizers designed for slow release and targeted delivery using urea-coated hydroxyapatite-NP significantly enhance the rice crop yields, even with half the amount of urea compared to conventional fertilizer. Specific natural polymers, such as chitosan NPs, have been used for the controlled release of N, P, and K by foliar uptake in wheat.

In biologically synthesized NPs, the presence of protein corona (PC), which is a highly dynamic and complex biomolecular coating around the NPs in physiological environments, dictates events such as the specific binding of the NP to the cell surface, its internalization, and its subsequent transport. Protein binding on the NP surface alters its size and surface charge, influencing NP uptake and biodistribution within the cell. These proteins control cellular receptors, which aid the protein-NP complex in cellular internalization and subsequent transport and assimilation. Such protein coatings can facilitate the internalization of NPs coated with specific proteins for a target cell type or tissue and/or can be masked (PEGylation) to prevent recognition by the immune system, ensuring prolonged circulation in the host cells without being recognized as a potential threat. This might provide broader avenues for developing NP-based targeted actions with minimized non-specific uptake. How proteins are adsorbed and arranged on the NP surface influences the movement and bioactivity of the NPs at the cellular level. The protein-NP complexation modifies the hydrodynamic, plasmonic, and electrokinetic properties of the NPs. Numerous investigations have revealed significant connections between PC composition and NP-physicochemical properties, giving biological character and identity to the NPs. In a plant model, protein coating of NPs can impart targeted delivery features. Recently, gold-NPs coated with antibodies for α-1,5-arabinan were accurately targeted to stomata in *Vicia faba*, and gold-NPs coated with bovine serum albumin (BSA) protein had a specific affinity to the trichomes. In a biological system, various bio-physicochemical properties, such as ionic strength, temperature, pH, and density of PC, impact NP function, critically characterize the physiological interactions between NP and the cell, and provide suitable modifications of cellular functions in the plant system. In addition, various factors are also involved in affecting the formation of PC surrounding the NPs. The protein adsorption on NP is a multifactorial process that relies on the amount of accessible protein, the surface area accessible on NP, and the protein's affinity. This is a competitive process of protein adsorption on the NP surface and is subject to its affinity, duration of contact, and abundance; it is known as the "Vroman effect." Dissemination of NPs and the development of PC inside the cells during bio-dispersion and translocation to different locations within the cells are controlled by the Vroman effect. Such time-based protein kinetic studies play an important role in identifying the key protein players in the hard and soft coronas, as well as their residence times and association dissociation rates. The surface charge impacts the density of proteins adsorbed but does not change the profile of the proteins adsorbed. Any communication driven electrostatically between NP and protein is a functional component of distinct local charge districts on the surface of the protein, but not the mass isoelectric point of the protein moiety itself. Protein binding to the NP in a unique orientation is reversible and free of modification in the protein surface charge. Studies carried out in this context have shown that PC composition is characterized by the size of the NP; larger NPs tend to adsorb proteins of higher molecular weight than smaller ones. Likewise, the size of the NP–PC complex fluctuates with the surrounding media. The NP surface curvature assumes a significant role in the adsorption of proteins, administering the structure and composition of the PC. Several studies have shown that fewer changes occur in the structure of the proteins adsorbed on curved surfaces of NP than on flat surfaces. The PC profile changes quantitatively and qualitatively with exposure or incubation time; increases in the concentration of proteins adsorbed influence NP-uptakes and cell interactions. The PC is very dynamic for an initial couple of minutes or hours upon interaction with NP in a biological medium. The underlying period of "soft corona" development might be rapid, but the gradual formation of "hard" corona is a drawn-out procedure that changes the physiological reaction of the NP-PC complex depending on the phase of evolution and structure of PC at that very moment. PC surrounding NPs provides stability against broad homoaggregation, which is additionally governed by the media composition and chemistry of the NP surface. As a result, in biological media, the surface properties of NPs deviate from their initial design, and they should be engineered with this in mind to enable communication with cell membranes and receptors, as well as checking for the impact of proteins bound to NPs on their uptake, circulation, and assimilation in bio-systems (Camara et al. 2019).

8.4 Ethical and Safety Issues

Ethical and safety issues related to the use of NPs for agronomic applications must be considered and carefully evaluated. Granting the presence of possible nanoproducts in the food and agribusiness domains, it is a widely acknowledged fact that sparse, reliable information is available concerning the safe use of NPs in light of the lack of safety assessment regulations. A robust life cycle assessment (LCA) is essential to thoroughly assess the environmental impacts of the nanofertilizers and to design suitable fertilizer products and appropriate dosages for their effective uses. Toxicity and risk assessment studies will identify the risks associated with the spread and bioaccumulation of NPs in the food chain and will enable us to make sound policy decisions for their broader commercialization and influence consumer decisions directly. Biocompatibility studies are needed for the successful commercial applications of NPs. Biocompatibility is primarily attributed to the protein coating surrounding the NP that interacts with the cellular components and machinery of the test organism. In addition to the role of the protein coating, the metal core in the cell subjected to metal homeostasis can explain specificity and elicit cellular reactions in response to NP-invasions. The highly specific areas of NPs interact with the cells and govern NP-uptake and intracellular localization. The agglomeration of NPs can alter some surface properties, leading to different cell interactions, changing the surface area available for protein adsorption, and causing conformational changes in the proteins, triggering differential uptake and localization patterns. Thus, controlling agglomeration is important for enhanced bioavailability to limit spread and improve safety. In addition to the surface area and protein coating, the net change in amino acid residues determines the preferential active uptake of NPs. Thus, successful commercial application of NPs requires nano-toxicological and risk assessment studies, which form the baseline for the development of safe NPs.

It is generally perceived that nanoform is more toxic than its bulk form counterpart, but this needs thorough investigation backed by toxicological evidence. Exposure and risk assessments depend on examinations related to the environmental fate of a compound. Some of the investigations have examined nano-agrochemicals regarding their implications for the environment, with the results insufficient for a reliable evaluation of the risks related to their uses. Comprehensive risk assessment and biocompatibility studies (cyto- and phytotoxicity) need to be prioritized for NP acceptance and to identify the fate of NPs in the soil-plant system for their successfully acceptable applications in the field of agriculture. Expanding regulatory pressure, joint efforts across the fields (e.g., industry, research, and regulators), and the integration of scientific disciplines with social development and law will ensure the development of practical legal frameworks offering consumer and public acceptance. A transdisciplinary approach underpinned by the One Health concept is needed to support the sustainable development of these technologies and the safe use of nanotechnology. This concept is endorsed by several organizations, including the World Health Organization (WHO) and the Food and Agriculture Organization (FAO), and is defined by the FAO. The One Health vision is a unifying force to safeguard human and animal health. Hence, the use of nanobiotechnology in agriculture requires a deliberate effort to overcome the difficulties presented by the regulations and policies that have not yet been put in place or are not fit for the purpose as discussed (Camara et al. 2019).

8.5 Stimuli-Responsive Systems

Before considering the targeted deliveries of nanopesticides for agricultural applications, it will be useful to have a quick overview of the connected developments in the area of targeted deliveries of disease-specific molecular species of pharmaceuticals that are currently being considered as drug nanocarriers and are expected to target disease-specific areas for improving the efficacy often altered by nonspecific cell and tissue biodistribution. A broad range of nanocarriers with varying sizes, architectures, and surface properties have, of late, been designed using liposomes, polymer NPs, micelles, dendrimers, and inorganic NPs made of iron oxide, quantum dots, gold, or metal oxide frameworks, which are extremely small to allow systemic (intravenous) or local (mucosal) administration for promoting their diffusion within the cells of living organisms. Moreover, currently available surface functionalization methods impart nanocarriers the ability to control, at least in part, their pharmacokinetics and biodistribution, as noted in the PEGylation-induced steric repulsion of antibodies, which enhance phagocytosis and endow them with *in vivo* longevity and the capability of extravasation through the endothelium of inflammatory tissues. Functionalization with biologically active ligand molecules helps in the targeting of specific cells. Nanotechnology-based targeted delivery has shown promising results in preclinical animal models. However, the translation of both the enhanced permeability and retention combined with ligand recognition into the clinic still remains questionable as a consequence of the stochastic nature of ligand-receptor interactions and the difficulties in controlling the release of the drug from targeting nanocarriers. Diffusion-induced leakage of the drug being non-specific to cells, tissues, or organs, one needs more efficient delivery strategies. One option is to use on-demand processes known as "switch on/off," which allow for precise spatial, temporal, and dosage control. Accordingly, on-demand delivery is becoming accessible using stimuli-responsive systems that not only recognize the microenvironment but also react, mimicking the responsiveness of living organisms. The requirement of biocompatible materials that should undergo protonation, a hydrolytic cleavage, or molecular or supramolecular conformational changes in response to a desired stimulus or that are susceptible to specific physical stimulation The stimuli-responsive delivery concept was proposed in the late 1970s with the use of thermosensitive liposomes for local release of drugs through hyperthermia. More research has been done on stimuli-responsive materials for drug delivery over the last decade, particularly on their design and application as nanocarriers. Stimuli-responsive nanodevices may be sensitive to specific endogenous stimuli, such as a lowered pH, a higher glutathione concentration, or an increased level of certain enzymes like metalloproteinases. At the cellular level, pH sensitivity can either trigger the release of the transported drug into late endosomes or lysosomes or promote the escape of the nanocarriers from the lysosomes to the cell cytoplasm. At the tissue level, one can take advantage of microenvironmental changes associated with neoplastic diseases as well as pathological situations such as ischemia, inflammatory diseases, or infections. Extracorporeal physical stimuli can also be applied. For instance, the targeted delivery of pharmacologically active molecules to a diseased area in the body can be magnetically guided by using ultrasmall iron oxide-based NPs. Sustained release can also be achieved by thermo-, photo-, or ultrasound-sensitive nanoparticulate systems (Mura et al. 2013).

8.5.1 pH Responsive Nanocarriers

Nanocarrier systems responding to the pH-changes deploy the ionizable functional groups like pyrimidines, carboxylates, amines, phosphates, and sulfonates that present charges at certain pH values subjected to electrostatic interactions. The phenomena, like swelling or shrinking of the macromolecules as a result of pH changes, are accordingly put to use. There are essentially two types of pH-sensitive materials: those having acidic groups in the structure (i.e., –COOH and $-SO_3H$) and those having basic groups ($-NH_2$), which swell at basic and acidic pH, respectively.

Biopolymers possessing desirable biocompatibility, biodegradability, and nontoxicity, are the required materials for the development of nanocarrier systems used in eco-friendly formulations that are ideal for the formation of nanocarriers responding to pH stimulus.

Out of several routes for pH-responsive nanocarriers, encapsulating a pesticide molecule offers an alternative route to using ionic or covalent interactions or trapping in a polymeric matrix within the system during NP-formation. A classic example uses alginate and chitosan. The alginate polymers with carboxylic groups are deprotonated at basic pH, possessing a negative charge. Chitosan has amine groups that are protonated and acquire positively charged functional groups at an acidic pH. Using ionic-gelation of alginate and chitosan biopolymers for preparing nanocarriers in a solution of pH ~4.7, the carboxyl groups are attracted to the amine groups, forming cross-links between the polymers and imparting pH-responsive changes. Consequently, the protonation or deprotonation of the polymers weakens or strengthens the electrostatic interactions, resulting in an increase or decrease in the release of a chemical compound. This system might be used only in biopolymers to control the opening of the polymer mesh, hence affecting the release of the active agent.

Other non-polymeric nanocarrier systems might deploy pH-responsive biopolymer coatings to prevent a release burst. Another approach is to functionalize the surface of the NPs with other compounds linked by electrostatic interactions in order to induce a response mechanism modulated by pH changes. For example, silica-NPs offer very quick initial release of the incorporated ingredients, which could limit their applications. However, surfaces rich in hydroxyl groups permit the functionalization or coating of these carriers. The pH-responsive nanocarriers are incidentally thermosensitive (Camara et al. 2019).

8.5.2 *Temperature-Responsive Nanocarriers*

Such systems are well-suited for agriculture due to the constantly changing temperature in the environment, which directly affects the emergence and development of agricultural pests such as insects, weeds, and fungi. At low temperatures, temperature-dependent water-soluble polymers have been extensively studied for their thermosensitive feature for controlled release. High-viscosity water-soluble carriers are widely employed to achieve the controlled release of active ingredients, as observed in the case of PEO-PPO-PEO block copolymers that are sold as Poloxamers® and Tetronics®. These polymers exhibit sol-gel phase transitions below or near physiological body temperatures. Low concentrations solutions in a liquid are known to form monomolecular micelles whereas at higher concentrations multimolecular aggregates are formed. Above the critical gelation temperature, the micelles are arranged in a crystalline structure, hence forming hydrogels (Vyas et al. 2018).

8.5.3 Redox-Responsive Systems

In redox-responsive systems, the stimuli causing the release of the active compounds use endogenous factors of antioxidant molecules like glutathione. Redox-responsive systems were studied first of all in the medical area, where they were used for the release of chemotherapeutic agents. Tumour cells with a higher concentration of reducing agents trigger the release due to the cleavage of a disulphide bond in the presence of the reducing agent. As a result, it is necessary to functionalize the nanocarrier with thiol-groups that are linked to the gatekeeper via disulphide bonds, which are easily cleaved. Several redox-responsive systems were developed using gatekeepers including organic components, inorganic crystals, polymers, and biomolecules. Some examples of gatekeepers are cyclodextrins, curcumin, ZnO QDs, PEG, and hyaluronic acid (Frickenstein et al. 2021).

8.5.4 Enzyme-Responsive Nanocarrier Systems

Enzyme-responsive materials are finding increasing applications in drug delivery due to their higher specificity and selectivity in response to internal biological stimuli. However, studies using enzyme-responsive materials for the delivery of agrochemicals have just started coming up. Exploiting enzymes as triggers for the smart delivery of agrochemicals has many advantages, such as specificity, accuracy, and efficiency, involving precise chemical reactions that occur under mild environmental conditions. Enzyme-sensitive polymers interact with their biological surroundings, resulting in appropriate responses. The development of NMs deploys the expression level of specific enzymes, resulting in an enzyme-mediated response leading to the controlled release of active compounds in a site-specific manner. Many enzymes are used as triggers for the release of active ingredients in order to achieve efficient pest control, especially in the case of enzymes present in the salivary glands and mid-gut of larvae and insects in the soil and produced by phytopathogenic fungi involving carbohydrases and proteases. The soil enzymes like urease, alkaline phosphatase, dehydrogenase, and catalase have been proposed besides those involving phytopathogenic fungi release enzymes like pectinases and cellulases, which are responsible for the degradation of plant cell walls (Song and Choi 2022).

8.5.5 Photo-Responsive Nanocarriers

Photo-responsive NPs allow spatiotemporally controlled release of active molecules using light irradiation of different wavelengths in the UV, visible, and IR ranges. Polymeric properties that change due to light stimulus include polarity, charge, conjugation, conformation, amphiphilicity, and optical chirality, among others. Such changes result in macroscopic changes in the properties of the polymers, such as shape, wettability, adhesion, optical properties, conductivity, solubility, and other similar ones. The release of active compounds from these photo-responsive polymers starts instantaneously after structural modification of the nanocarrier by irradiation at a specific wavelength. Light-controlled carriers are obtained by incorporating photoactive molecules like azobenzene, ortho-nitrobenzyl, coumarin, and spyropyrane into polymer matrices to stimulate the release of the enclosed compounds, which can occur by either triggering degradation of the polymer or a linker between the polymer and a small molecule, or by affecting changes in the polarity of the polymer. Coumarin and o-nitrobenzyl groups containing polymers are cleaved into smaller molecules, while the structures of azobenzene and spyropyrane polymers are induced reversibly in the presence of certain wavelengths of light. Light-controlled release can also act as an "on/off system," where the presence of light releases the active compounds, and in the absence of light or at a different light frequency, the system is reversed with release switched off (Camara et al. 2019; Pan et al. 2021).

8.5.6 *Stimuli-Responsive Systems: Agricultural Applications*

Stimuli-responsive smart delivery systems are expected to add to the goal of achieving sustainable agricultural practices. Stimuli-responsive nanoformulations release their active ingredients in response to environmental triggers and biological demands more precisely by means of targeted delivery or controlled release mechanisms. A variety of factors act to restrict crop growth and production, including biotic stress caused by insects, pathogenic diseases, nematodes, and weed competitors, as well as abiotic factors such as extreme temperatures, poor soils, the unavailability or poor uptake of nutrients, metal-contaminated soils, drought, flooding, and high salinity. Therefore, stimuli-responsive materials have been examined for the release of agrochemical compounds in response to pH, temperature, redox conditions, enzymes, and light (Camara et al. 2019).

8.6 Chapter Summary

The application of nanotechnology in agriculture is still in its infancy. The lower profit margins in agriculture compared with other sectors like biomedicine have resulted in fewer sound investments and research ventures that support nanotechnology in agriculture, both for producers and consumers. Joint efforts between disciplines at each phase of the development and assessment of agrochemicals (e.g., plant material, formulations, and ecological researchers) will advance nanoagriproducts by overcoming numerous limitations in the agrochemical sector and will add value to the existing products. On the research front, inconsistent results on the mechanism of NP biosynthesis, its cellular uptake and mobilization in the plant system, and factorial analyses of their connection with biological systems have increased the need for comprehensive understanding in this area to design safe and effective NPs. It is critical to recognize and fill the knowledge gap concerning the fate and effect of NPs as it relates to precise and effective exposure and risk assessment for a sustainable environment. A fair and reasonable evaluation of nano-agricultural inputs should focus on assessing both the challenges and opportunities related to their use, relative to the current solutions.

The pesticide field requires continued research for the development of improved environmentally responsive, targeted, controlled-release pesticide formulations. Effective systems that respond to internal bio-stimulation pose a significant challenge in this regard.

Although stimuli-responsive properties have the potential to reduce pesticide premature degradation, improve efficacy, and reduce collateral effects on nontarget organisms, there are numerous drawbacks that limit their widespread application.

For safer uses of nanopesticides, the development of nanoformulations based on green nanotechnology is imperative to offer low-cost, simpler procedures, and controlled-release features, ultimately leading to smart delivery nanopesticides for sustainable agricultural development.

Further Reading

Ahmad F, Ahmad S (2021) Biosynthetic route for the functionalization of nanomaterials. In: Kumar V, Guleria P, Dasgupta N (eds) Functionalized nanomaterials II – applications. CRC Press, Boca Raton. https://doi.org/10.1201/9781351021388

Camara MC, Campos EVR, Monteiro RA et al (2019) Development of stimuli-responsive nano-based pesticides: emerging opportunities for agriculture. J Nanobiotechnol 17:100. https://doi.org/10.1186/s12951-019-0533-8

Frickenstein AN, Hagood JM, Britten CN et al (2021) Mesoporous silica nanoparticles: properties and strategies for enhancing clinical effect. Pharmaceutics 13. https://doi.org/10.3390/pharmaceutics13040570

Mura S, Nicolas J, Couvreur P (2013) Stimuli-responsive nanocarriers for drug delivery. Nat Mater 12:991–1003. https://doi.org/10.1038/nmat3776

Pan H, Li W, Wu L et al (2021) β-cyclodextrin-modified mesoporous silica nanoparticles with photo-responsive gatekeepers for controlled release of hexaconazole. Coatings 11. https://doi.org/10.3390/coatings11121489

Song SJ, Choi JS (2022) Enzyme-responsive amphiphilic peptide nanoparticles for biocompatible and efficient drug delivery. Pharmaceutics 14. https://doi.org/10.3390/pharmaceutics14010143

Vyas B, Pillai SA, Bahadur A et al (2018) A comparative study on micellar and solubilizing behaviour of three Eo-Po based star block copolymers varying in hydrophobicity and their application for the *in vitro* release of anticancer drugs. Polymers 10. https://doi.org/10.3390/polym10010076

Chapter 9
Nanopesticides in Agriculture: Some Examples

9.1 Introduction

The development of stimuli-responsive smart delivery (targeted delivery) systems is quickly becoming the driving force behind achieving sustainable agricultural practices. Stimuli-responsive nanoformulations are progressively becoming capable of releasing their active ingredients (AIs) in response to the triggers applied externally to meet the crop demands more precisely at the right time. A variety of other factors also act to restrict crop growth and production, including biotic stress caused by insects, pathogenic diseases, nematodes, and weed competitors, as well as abiotic factors such as extreme temperatures, poor soils, poor uptake of nutrients, metal-contaminated soils, drought, flooding, and high salinity. Therefore, stimuli-responsive materials have been investigated for the release of agrochemical compounds in response to pH, temperature, redox conditions, enzymes, and light. A number of examples have already been reported in this context, as briefly described below.

These observations made by different research teams slowly converge on the situation that, in the event that the underlying mechanisms of interactions between nanoparticulate systems and soil, including plants, are better understood, more useful steps will be accomplished in the direction of sustainable agriculture without harming the ecosystem as well as human populations and livestock.

9.2 Insect and Nematode Control

The development of transgenic crops has stirred public concern about impacts on the ecological services of soil fauna, such as soil nematodes. Few investigations examined whether the cultivation of transgenic rice would pose greater threats to

A. Nazeer et al., *Targeted Delivery of Nanopesticides and Nanofertilizers in Sustainable Agricultural Farming*, Nanotechnology in the Life Sciences,
https://doi.org/10.1007/978-3-031-41333-9_9

soil nematodes and associated ecological functions than the insecticide application. The determinants of the soil nematode community in the paddy fields being unclear, a 3-year field study was conducted using rhizosphere soil samples of transgenic-Bt rice against its counterpart, non-Bt parental rice, and non-Bt rice with insecticide application. It was proposed that the effects of pest management practices on soil nematode abundance and metabolic footprint changed with trophic group and sampling time. Significantly different structure and composition of the soil nematode community were noted in the three treatments examined. The effects of pest management practices on nematode abundance and metabolic footprints depend on trophic group and sampling time. However, pest management practices exerted no apparent effect on nematode diversity and community composition. Soil nutrient availability and the C:N molar ratio are the primary regulating factors of the soil nematode community in rice paddy fields. The changes in abundance, diversity, and metabolic footprints associated with the crop growth stage outweighed the application of Bt rice and insecticides. Over three years of rice cropping, the cultivation of Bt rice Huahui-1 had no discernible negative impact on the soil nematode community in the rhizosphere soil (Yang et al. 2020).

9.2.1 pH-Responsive Examples

The basic mechanism involving reactions of ingested nanocarriers inside the intestines of agricultural pests like insects and nematodes with a range of acidic and alkaline pH conditions has been used in activating the release mechanism of active ingredients. For example, an alginate nanocarrier system was reported to control the insecticide cypermethrin. The acidic pH led to a high degree of ionization of the alginate polymer, causing strong interaction with calcium ions that increased the cross-linking of the polymer mesh, hence making the interior of the NPs more hydrophobic and reducing release of the insecticide. With increasing pH, the electrostatic interactions diminished, allowing the entry of water into the system and the consequent release of the active agent.

Other pH-responsive systems were developed for insecticides, comprising silica-NPs coated and functionalized with polystyrene and (trimethoxysilyl)propyl methacrylate for the delivery of the insecticide abamectin. They showed that after 15 days at pH 5 and 7, there were 14–15% releases, demonstrating that the system prevented premature release of the insecticide, whereas at pH 10, the release of the insecticide reached 39% after 5 h and 87% after 15 days.

Another study revealed properties of cyclodextrin-silica-NPs containing avermectin stimulated by pH-responsive enzymes (amylase). At alkaline pH, there was a 40% release after 17 days, while a release of ~6% was obtained at neutral and acid pH over the same period.

Another pH-responsive release of avermectin system was proposed with a poly(succinate) containing NPs showing release percentages of 34, 62, and 85% at pH 5.5, 7, and 8.5, respectively. In this case, the response was related to the collapse of the NPs at a higher pH, consequently releasing the insecticide.

Systems involving alginate and chitosan combined with acetamiprid showed 50% of the insecticide release after 24 h at pH 10, while the same amount was released after 36 h at pH 7 and 4.

Similarly, pH-responsive systems have been used for other pests that damage agricultural crops, such as nematodes. For example, a NP-system consisting of γ-polyglutamic acid and chitosan was reported for the transport of avermectin, used as a nematicide. The system showed an initial pH-independent release of 20%, while subsequently there was a release of 69.5% at pH 8.5, compared to 60.4% at pH 7 and 57.5% at pH 5.5. The nanocarrier system showed high stability at pH 5.5, due to the electrostatic interactions between chitosan and the carboxyl groups of γ-polyglutamic acid. As the pH increased, this interaction diminished, subsequently increasing the release of the nematicide. In agricultural systems, pH variations occur in soils, in different plant organs, during physiological processes of the plant, and in the maturation of fruits, as well as due to the presence of pathogens or agricultural pests. Therefore, the fact that nanocarrier systems are able to release active agents in response to pH changes represents a major strategy for the development of new nanopesticides since the release kinetics are modulated according to pH changes (Hernández-Soto and Chacón-Cerdas 2021).

9.2.2 Temperature-Responsive Systems

The pyrethrin-based insecticide employed temperature-responsive nanocarriers consisting of nanomycetes. The release was evaluated using three temperature conditions, namely, 13, 18, and 26 °C, by simulating the temperatures of the mosquito breeding stages in water. The system released 12% of the pyrethrin in the first 16 hours at low temperatures, with no further changes after that. At higher temperatures, the release increased over time, reaching 31.9% at 18 °C and 49.7% at 26 °C. Chitosan-coated silica-NPs showed higher release of avermectin by raising the temperature, with releases of 18.82% at 25 °C and 34.21% at 50 °C, demonstrating that coating with chitosan was a better option for altering the temperature-sensitive release profiles. The greater release at a higher temperature was due to changes in the thermodynamic movement of the molecules, which increased the diffusion of the pesticide molecules through the chitosan.

9.2.3 Enzyme-Responsive Example

Enzyme-responsive polymers have been studied for the release of insecticides in the presence of enzymes found in the guts of insects and larvae. After the synthesis of microcapsules containing emmamectin benzoate using silica cross-linked with epichlorohydrin-modified carboxymethylcellulose, the release behavior of these microcapsules was evaluated in the presence of cellulase, which cleaved the cellulose wall material into smaller fragments, releasing the active ingredient. Almost 30% of the emmamectin was released after 1 h of cellulase exposure, with ~80% release reaching after 30 h. In the absence of enzymes, < 20% was released after 30 h. Use of the silica-epichlorohydrin-carboxymethylcellulose microcapsules reduced the genotoxic effects of the active agent and enhanced the insecticidal activity against *Myzus persicae*, compared to the use of emmamectin benzoate alone.

The application of an α-amylase responsive carrier was reported for the development of a smart delivery system that responded to internal stimuli produced by insects, triggering biocide release. The developed system used α-cyclodextrin anchored on hollow mesoporous silica loaded with avermectin. The NPs showed a strong response to the presence of α-amylase, with avermectin release of 41.64% up to day 17, compared with only 5.88% release in the absence of α-amylase, during the same period. The mortality of larvae remained >80%, even after 14 days, while the effectiveness of a commercial avermectin formulation decreased to 43.4%. The avermectin nanoformulations remained active 44 days after application, when 6.63% larval mortality was observed (Hou et al. 2021).

9.2.4 Photo-Responsive Examples

Coumarin was noted as a photo-trigger for insecticide delivery, loaded with the advantageous properties of coumarin to improve the stability of molecules, together with pesticidal activity, strong fluorescence, fast release rates, a broad absorption wavelength range, and high biocompatibility. A photocaged insecticide composed of coumarin and fpronil was reported, with a strong covalent link between fpronil and coumarin resulting in an on/off system with light-dependent insecticide release. The system was able to provide efficient control of *Aedes* larvae when exposed to blue light and sunlight.

In a similar study, caged insecticides using spirotetramatenol and coumarin were reported with covalent linking. The photo-triggered release profile was evaluated by exposing the system to blue light (420 nm) and sunlight, which showed efficient release of spirotetramatenol using both irradiation sources. In the dark, the caged insecticide showed low insecticidal activity against *Aphis craccivora*, while blue light irradiation significantly increased mortality.

In another study, a sex pheromone was caged as an insect attractant using different photo-removable protecting groups, including 7-hydroxy-4-hydroxymethylcoumarin, 1-pyrenemethanol, 9-anthracenemethanol, and 2-(hydroxymethyl) anthraquinone. All the photoactive molecules were able to release the pheromone, with UV (≥350 nm) or sunlight irradiation being essential for the controlled release of (Z)-11-hexadecen-1-ol. The use of infrared-responsive nanocomposites was reported to prevent the loss of hydrophilic pesticides and improve their efficiency. For this, graphene oxide (GO) was loaded with hymexazol (Hy) by absorption, followed by coating the GO surface with polydopamine (PDA). The Hy-GO@PDA nanocomposite responded to photothermal heating using a near-infrared (NIR) laser at 808 nm. The release behaviour showed that the Hy-GO@PDA was responsive to both pH and NIR, so the release of Hy could be controlled by either pH or NIR irradiation. The highest Hy release was 75% after 120 h at pH 9.0 under NIR irradiation.

Photo-responsive formulations have broad applications in the agroindustry, where the development of new, cost-effective sunlight-triggered delivery systems is highly desirable for the encapsulation of fertilizers, pesticides, and plant growth regulators.

9.3 Control of Phytopathogens

An extensive study was carried out on phytopathogenic members of the *Sclerotinia* genus that cause widespread disease across a broad range of economically important crops. Particularly, *Sclerotinia sclerotiorum* was considered one of the most destructive and cosmopolitan of plant pathogens. The epidemiology of this pathogen, its economic impact on agricultural production, and measures employed to control the disease were reviewed using broad approaches to tackle *Sclerotinia* diseases, which include cultural practices, crop genetic resistance, chemical fungicides, and biological controls (O'Sullivan et al. 2021).

Huge crop losses occurring across a broad range of horticultural and farming sectors have pressed for a new and effective control to be developed against *Sclerotinia* pathogens by exploring newer fungicides with a wider window of preventative activity. A better understanding of the processes involved would not only help in the optimization of the formulation and delivery but also promote screening for more efficient or potent strains of commercial value. Moreover, combinations of chemical fungicides and biocontrols, or multiple biocontrols, would enable the development of multiple modes of action and/or synergistic effects. Such approaches, once incorporated with farming cultural practices, would lead to a coordinated attack on *Sclerotinia*. There are possibilities of using genetic modification of certain crops to attain complete disease resistance against *Sclerotinia* (O'Sullivan et al. 2021). Further studies conducted in this direction are highlighted below.

9.3.1 pH-Responsive Example

Biodegradable hydrophilic biopolymers that hydrolyze at alkaline pH were used to create systems for controlling phytopathogens based on poly(succinate). A poly (succinate) nanocarrier system was subsequently reported for the disease control of plant phloem. The pH of most plant tissues is acidic in contrast to alkaline vascular phloem tissues. Accordingly, this type of nanocarrier system could be used against pathogens such as citrus huanglongbing (known as "greening"), which specifically infects the phloem, leading to plant death.

The pH-responsive systems could also be used to prevent the contamination of fruits by pathogens, which could reduce yields by up to 30%. Contamination of fruits by fungi results in a lowering of pH due to the production of enzymes such as esterases and pectinases. Fruit ripening is also accompanied by an increasingly acidic pH. Nanocarrier systems were, accordingly, developed to control the growth of contaminant organisms in food, as well as to avoid early maturation. For example, silica-NPs coated with chitosan were reported to deliver the fungicide prochloraz. In the absence of the esterase enzyme and at pH 7, only 18.9% of the fungicide was released for 30 days. However, a decrease in pH and the presence of the enzyme stimulated the system, with 82% release obtained during the same period.

Chitosan protonation at an acidic pH is found to expose the pores of the silica-NPs and consequently enable the release of the active agent. A system of GO-NPs was also reported to release salicylaldehyde, where the binding of salicylaldehyde to the GO-NPs was mediated by the hydrazone molecule. At an acidic pH, 50% of the active agents were released after 50 h, compared to only 15% at a neutral pH. The release was due to the hydrolysis of hydrazone at an acidic pH, releasing the active agent and reducing earlier ripening.

Silica nanoparticles and curcumin in chitosan gel could produce an intelligent food packaging system for preventing the growth of harmful organisms. A faster release was observed under acidic conditions compared to neutral conditions. At an acidic pH, chitosan being protonated results in electrostatic repulsion between the polymer chains, hence increasing the volume of the polymer mesh and releasing the active agent, in contrast to a neutral pH where chitosan is in a neutral state (Beyene et al. 2021).

9.3.2 Redox-Responsive Examples

A redox-responsive system was developed for the plant growth regulator salicylic acid, which is a hormone that regulates plant defence mechanisms in the presence of biotrophic pathogens. Silica-NPs were functionalized with thiol groups, followed by coating with decanediol, forming specific disulphide bonds that were cleaved in the presence of glutathione. In *in vitro* release assays, an increase was reported in glutathione levels, causing increased release of salicylic acid. Plant assays using *Arabidopsis* as a biological model for evaluating the expression of the PR-1 defence gene were found to be mediated by salicylic acid. An increase in glutathione concentration in the plants resulted in the nanocarrier system starting to release salicylic acid, which induced high expression of the PR-1 gene and could increase the resistance of plants against pests.

9.3.3 *Temperature-Responsive Example*

Silica-NPs containing the fungicide thymol and functionalized with OH, NH_2, and COOH groups released more of the fungicide, whereas NPs functionalized with hydroxyl and carboxyl groups at elevated temperatures released more AIs than those functionalized with amino groups.

9.3.4 *Enzyme-Responsive Examples*

Pectinases from phytopathogenic fungi were examined for triggering the release of antibiotics in order to control plant diseases. The synthesis of kasugamycin conjugated with modified pectin by amide bonds, when evaluated, showed triggered release in the presence of *Pseudomonas syringae* pv. *lachrymans*. A sustained-release profile was noted in the case of kasugamycin-pectin, which was controlled by the pectin degradation enzymes produced by *P. syringae*. The kasugamycin release rate increased with time, indicating that greater quantities of pectinase were released during microbial growth, resulting in faster release of the active compound.

In another study, the preparation of enzyme-responsive kasugamycin microcapsules was reported using mesoporous silica-NPs conjugated with pectin by the addition of disulphide bonds. Dual-responsive kasugamycin microcapsules were developed with protective effects against degradation by temperature and light. The delivery system was triggered by either pectinase or glutathione stimuli, with the release of kasugamycin reaching 88.13% after 30 h in the presence of both stimulants, while values of 68.87% and 77.74% were obtained in the presence of pectinase or glutathione alone, respectively. The efficacy of kasugamycin microcapsules against *Erwinia carotovora* was lower during the first few days, subsequently increasing over time due to the microorganism's continued production of the triggers. It was concluded that the product could remain effective for approximately 3 weeks, indicating its great potential for use in agricultural applications (Peng et al. 2021).

9.4 Weed Controlling Nanocarriers

Yield losses in crops due to weeds depend on several factors, such as weed emergence time, weed density, type of weeds, crops, and the like. Uncontrolled weeds can result in 100% yield loss. In terms of yield losses, weeds amounted to 2.7 million tons of grain at a national level in Australia. In India, these costs were much higher. These studies from different economies indicate the substantial yield and economic loss caused by weeds.

The current total global grain production is ~2.1 billion metric tons. Assuming an overall yield loss of 10% due to weeds, the total loss in grain production is ~200 million metric tons. Reducing this loss by half, grain production would increase by 100 million metric tons, which could reduce hunger worldwide.

In developing countries, where farm sizes are small, weeds are removed manually. This practice is becoming less common as a result of labour migration to cities and rising wages in the agriculture sector. Hand-weeding is being replaced by herbicide use. In developed countries, such as Australia and the USA, herbicides are already widely used to control weeds. However, overreliance on herbicides with similar modes of action has resulted in the evolution of herbicide-resistant weeds. Presently more than 500 unique cases of herbicide-resistant weeds have been reported globally. Out of this total number of cases, more than 160 are from the USA and over 90 are from Australia, making them the two countries with the highest number of cases of herbicide resistance. These nations are followed by Canada, China, and Brazil. The maximum number of herbicide-resistant weed species reported in different crops is in the order of: wheat > maize > rice (*Oryza sativa* L.) > soybeans > spring barley (*Hordeum vulgare* L.) > canola (*Brassica napus* L.) > cotton (*Gossypium hirsutum* L.).

New herbicides with different modes of action are, therefore, needed to manage herbicide-resistant weeds; however, no major mode of action has been introduced in the past three decades. These issues suggest the need to develop different weed management options and to consider the potential for integrating them with herbicide use. These concerns have also encouraged weed scientists around the world to develop ecologically-based weed management tools (Chauhan 2020).

9.4.1 Temperature-Responsive Nanocarriers

The herbicide glyphosate was proposed, involving a nanocomposite composed of attapulgite, NH_4HCO_3, amino silicone oil, and polyvinyl alcohol. An increase in temperature caused pores to form in the nanocomposite due to the dissolution of the poly(vinyl alcohol) surfactant, facilitating herbicide release. After 72 h, the amounts released were 13% at 25 °C, 22% at 40 °C, and 33% at 50 °C (Chen et al. 2018).

In the case of silica nanoparticles, the temperature affected the diffusion of the guest compound through the silica pores. Strategies such as functionalization with different functional groups or coating with polymers could be used to modulate the release of NPs at different temperatures. For example, release of the herbicide 2,4-dichlorophenoxy was reported from silica-NPs increased at higher temperatures, due to enhanced diffusion of the herbicide through the NP-pores to the external medium. The amounts of the active agent released after 15 hours were 96, 75, and 52% at temperatures of 40, 30, and 20 °C, respectively.

9.4.2 Enzyme-Responsive Nanocarriers

Soil enzymes are useful in triggering the release of bioactive compounds such as pesticides and fertilizers. For example, urease was found effective in triggering the herbicide release from the silica-NPs. Urease-responsive herbicide microcapsules were prepared by conjugating pendimethalin (PEI) onto the surface of isocyanate-functionalized mesoporous silica-NPs by means of urea bonds, which provided sites for cleavage by urease. This system, denoted silica-IPTS-PEI, showed a cumulative release of 81.94% after 30 h, while a release of only around 10% was achieved in the absence of urease. In addition, the formulation improved the durability of PEI applied for weed control. Accordingly enzyme-responsive formulations offer considerable benefits. The active compounds will be mostly released only in the presence of enzymes, thereby increasing the efficiency and safety of the product while reducing the effects on nontarget organisms. Another important factor to consider is that the use of such responsive formulations could significantly reduce the amount of chemicals used when compared to conventional techniques (Cantarella et al. 2018).

9.4.3 Photo-Responsive Examples

Recent studies on developing a light-responsive controlled-release system reported using biochar, attapulgite, azobenzene, and amino silicon to encapsulate the herbicide glyphosate. Azobenzene was used as a light-stimulated trigger since it has trans and cis isomers that can be transformed into each other under UV or visible light, consequently promoting the release of the herbicide. Continuous release of glyphosate was obtained, which was more strongly induced by UV-Vis irradiation (~100% glyphosate release), compared to exposure to daylight or dark conditions (~80% release). This effect was attributed to the simultaneous trans–cis and cis–trans transformations of the azobenzene molecules, induced by UV–Vis irradiation, which did not occur in daylight or in the dark. The particles also showed high leaf surface adhesion due to the high viscosity of amino silicon, as well as stability under different pH conditions. Weed control efficiencies of 93.7 and 46.7% were achieved when the light-responsive particles were used in the presence and absence of UV-Vis light, respectively, confirming the stimulus provided by azobenzene in the presence of irradiation (Liu et al. 2019).

A photo-responsive supramolecular vesicle loaded with paraquat was prepared by the self-assembly between cucurbit [8] uril (CB[8]), paraquat, and an azobenzene derivative. The vesicles showed good stability properties when stored in the dark for 210 days, with a slight increase in particle diameter from 208.1 to 293.5 nm. The cumulative release ratio showed the UV responsiveness of paraquat-loaded vesicles; in the dark, only about 10% of the paraquat was released after 10 hours, but when exposed to UV light, the release of paraquat reached 90% after 24 minutes under continuous irradiation. The loaded herbicide showed a safer profile than free paraquat in cell and animal models and similar herbicidal activity when compared with free herbicide against *Festuca arundinacea*. The photoactive molecule 3-nitro-4-bromomethylbenzoic acid (NBA) was used as a phototrigger in the synthesis of a system consisting of polyethylene glycol (PEG) and the pesticide dichlorphenoxyacetic acid (2,4-d). Some of the salient features of these delivery systems have been discussed by Song et al. (2019).

The 2,4-d-NBA-PEG system could self-assemble into core-shell micelles, for which the release profile showed the controlled release of 2,4-d under exposure to sunlight (99.6% in 8 h) or UV irradiation (99.5% in 9 min), while no release was observed in the absence of irradiation. The photolabile 2-nitrobenzyl group conjugated with carboxymethyl chitosan could produce photo-responsive micelles for the delivery and controlled release of the hydrophobic herbicide diuron. The cross-linked micelles released the herbicide in response to UV irradiation and sunlight, indicating their potential for the delivery of pest inhibitors. Other work has investigated the caging of 2,4-d with coumarin as well as the synthesis of a coumarin copolymer based on acrylate and polyethylene glycol loaded with 2,4-d, which presented advantages including stability, release of 2,4-d controlled by UV light, and extended herbicidal activity (Wan et al. 2022).

9.5 Nanocarriers for Soil Applications

The quality, nutritional level, and physicochemical properties of soil play major roles in the growth and development of plants. Stimuli-responsive NPs can be applied to soils in order to provide targeted and slow releases of fertilizers, fungicides, herbicides, and plant growth regulators.

9.5.1 *Redox Responsive Examples*

A redox-responsive system was developed using the encapsulation of abscisic acid, a plant growth regulator responsible for regulating physiological changes in plants and increasing glutathione levels under stress conditions. Trials with *Arabidopsis* plants under hydric stress conditions revealed induced physiological changes, such as stomatal closure to avoid water loss and high expression of the stress-related gene AtGALK2. The plants treated with NPs exhibited a higher survival rate under hydric stress. A multi-responsive NP-system for the delivery of gibberellic acid (GA3) was recently described. The exogenous application of this plant growth regulator could alleviate plant stress caused by drought and salinity. GA3 was enclosed in hollow mesoporous silica (HMS) capped with Fe_3O_4-NPs functionalized with carboxylatopillar[5]arene ammonium (WP[5]A). The release system could be controlled by various stimuli, including alkaline or acidic pH, polyamines, and ultrasound. The stimulus provided by polyamine was due to competition between 1,4-butanediamine and pyridine for interaction with WP[5]A. The ultrasound stimulus could weaken and destroy the host–guest interactions, which could be restored by removing the ultrasound source (Hoang et al. 2022).

9.5.2 *Temperature Change Responsive Chitosan*

Chitosan biopolymers responsive to temperature changes were studied in the context of plant growth regulator releases. It was reported that the temperature could alter the viscosity and hydrophobicity of the polymers, thereby modulating the release of GA3, with a faster release when the temperature was increased to 30 °C compared to the release at 25 °C.

9.5.3 *Photo-Responsive System*

A photo-responsive system was developed in another work using the photoremovable protecting group of coumarin for the smart release of plant growth regulators like indoleacetic and naphthoxyacetic acids. The system showed controlled release of the plant growth promoters when exposed to sunlight, resulting in improved root and shoot growth of *Cicer arietinum*. Novel stimuli-responsive NPs were synthesized for controlled fertilizer release in response to anion exchange. With higher availability of ions in the soil, an anion-responsive nanosystem based on polyethylenimine-modified hollow mesoporous silica NPs was obtained by electrostatic interaction between the carrier and the fertilizer (selenium). The release of selenate was significantly stimulated by valence and the anion content, with maximum release of selenate (90% after 10 h) obtained using a 5 mM phosphate solution. This condition also improved plant growth by 60.7% compared to the control, indicating that selenate was successfully transformed into organic Se by the plant with enhanced efficiency and avoidance of losses (Hossain et al. 2021).

9.5.4 *Magnetic Stimuli Responsive Example*

Systems employing magnetic stimuli represent another promising technology that could be used for the delivery of hormones and biocides by remote control. A magnetic nanosystem was reported for drug delivery, while for agricultural purposes, studies of magnetic NPs have so far been focused on soil decontamination.

Functional magnetic NPs designed for use in agricultural applications are finding their uses in controlling the current climate changes and pollution linked to anthropogenic activities. There are examples of magnetic NPs employed as effective adsorbents for the removal of antibiotics, pesticides, and toxic metal ions from contaminated wastewater; soil fertility promoters, enhancing the uptake of nutrients in crop plants; and magneto-assisted soil restoration agents enabling the removal of toxic soil contaminants (Spanos et al. 2021).

9.6 Limitations and Challenges

Many publications have shown that nano-based pesticides have great potential in the field of pest management, offering advantages for the environment and human health compared to the use of conventional pesticides. Using stimulus-responsive release nanoformulations, the main issues associated with conventional pesticides could be overcome. Some key areas of stimulus-responsive systems require improvement before they can be successfully marketed.

One of the factors influencing this low level of commercialization is that the great majority of studies are undertaken in universities and research institutes or by small companies set up for specific agricultural purposes. At the same time, large companies possess many patents, with the number increasing annually. Recently developed nano-based products for the agricultural sector do not reach the market because large companies are accumulating patents and waiting for opportunities for future exploitation following the development of promising commercial products.

Another issue is related to better understanding the risks related to the use of nano-based pesticides, which are complex and poorly understood. It is already known that the transport, bioaccumulation, and rate of degradation of nano-based pesticides are different from those of conventional pesticides, and hence further work is needed to elucidate them better. There are limited data available concerning the effects of nano-based pesticides on crop health, soil biodiversity, nontarget organisms, and human health.

The development of nano-based pesticides, however, suffers from a lack of effective strategies to understand their mechanisms of action. It is very important to evaluate the fate of nano-based pesticides in the environment in order to elucidate the interactions of these materials with terrestrial and aquatic organisms. Besides the problems mentioned above, it should be highlighted that economic issues remain a major hindrance for the commercialization of nanopesticides. The initial costs involved in developing nanopesticides are high, with positive financial returns only possible if large quantities of these products were to be used in crop protection, which is not currently the case.

In addition, the lack of regulation is a major flaw in the expansion of nanotechnology in agriculture. Another factor that limits the commercialization of nano-based pesticides is the high cost associated with registering a new active compound. The most significant challenges that should be addressed in increasing the commercialization of nano-based pesticides are listed as follows:

- Developing standard methods for reliable risk-benefit analysis;
- Continued research into the interactions of nanoparticles with plants or pathogens;
- Development of a tracking system for nanopesticides in the environment in terms of their impacts on food security and human health;
- Evolving International Legislation to ensure the safe development and application of stimuli-responsive nanopesticides; and
- Establishing a global network for communication among the public and private organizations engaged in developing nanopesticides.

It is clear that further in-depth research will be essential to address these issues prior to the full commercialization of nano-based products. It is important that there should be interaction among all the potential stakeholders, including universities, research institutes, companies, governments, NGOs, and consumers.

9.7 Chapter Summary

Because of the recent rapid growth in nanotechnology application in agriculture, many commercial companies are now forming partnerships with the scientific community to develop feasible nanopesticides for agricultural farming. In agricultural nanotechnology, particularly the development of stimuli-responsive nanoformulations, it would be helpful to maintain the stability of the active ingredient under environmental conditions, deliver the active ingredient to the target, decrease its dispersal in the environment, and prolong its biological activity. Stimuli-responsive systems are although well developed in medicine, their applications in agriculture are still limited. Continued research in the pesticide field is also required for the development of improved environmentally responsive, targeted, controlled-release pesticide formulations. Effective systems that respond to internal bio-stimulation pose a significant challenge in this regard. Although stimuli-responsive properties could reduce pesticide premature degradation, improve efficacy, and reduce collateral effects on nontarget organisms, there are still many drawbacks that prevent their widespread use. To ensure the safe use of nanopesticides, it is essential to develop nanoformulations that are based on green nanotechnology, as that offers low cost, simple procedures, and controlled-release features.

It is thus concluded that using smart delivery nanopesticides is highly promising as an effective tool for sustainable agricultural development in the years to come.

Further Reading

Beyene AM, Moniruzzaman M, Karthikeyan A et al (2021) Curcumin nanoformulations with metal oxide nanomaterials for biomedical applications. Nanomaterials (Basel). 11. https://doi.org/10.3390/nano11020460

Cantarella H, Otto R, Soares JR et al (2018) Agronomic efficiency of NBPT as a urease inhibitor: a review. J Adv Res 13:19–27. https://doi.org/10.1016/j.jare.2018.05.008

Chauhan BS (2020) Grand challenges in weed management. Front Agron 1. https://doi.org/10.3389/fagro.2019.00003

Chen C, Zhang G, Dai Z et al (2018) Fabrication of light-responsively controlled-release herbicide using a nanocomposite. Chem Eng J 349:101–110. https://doi.org/10.1016/j.cej.2018.05.079

Hernández-Soto A, Chacón-Cerdas R (2021) RNAi crop protection advances. Int J Mol Sci 22. https://doi.org/10.3390/ijms222212148

Hoang AS, Cong HH, Shukanov VP et al (2022) Evaluation of metal nano-particles as growth promoters and fungi inhibitors for cereal crops. Chem Biol Technol Agric 9. https://doi.org/10.1186/s40538-021-00277-w

Hossain A, Skalicky M, Brestic M et al (2021) Selenium biofortification: roles, mechanisms, responses and prospects. Molecules 26. https://doi.org/10.3390/molecules26040881

Hou Q, Zhang H, Bao L et al (2021) NCs-delivered pesticides: a promising candidate in smart agriculture. Int J Mol Sci 22. https://doi.org/10.3390/ijms222313043

Liu B, Chen C, Wang R et al (2019) Near-infrared light-responsively controlled-release herbicide using biochar as a photothermal agent. ACS Sustain Chem Eng 7:14924–14932. https://doi.org/10.1021/acssuschemeng.9b03123

O'Sullivan CA, Belt K, Thatcher LF (2021) Tackling control of a cosmopolitan phytopathogen: *Sclerotinia*. Front Plant Sci 12. https://doi.org/10.3389/fpls.2021.707509

Peng Y, Li SJ, Yan J et al (2021) Research progress on phytopathogenic fungi and their role as biocontrol agents. Front Microbiol 12. https://doi.org/10.3389/fmicb.2021.670135

Song N, Lou XY, Ma L et al (2019) Supramolecular nanotheranostics based on pillarenes. Theranostics 9:3075–3093. https://doi.org/10.7150/thno.31858

Spanos A, Athanasiou K, Ioannou A et al (2021) Functionalized magnetic nanomaterials in agricultural applications. Nanomaterials (Basel) 11. https://doi.org/10.3390/nano11113106

Wan J, Fan B, Thang SH (2022) RAFT-mediated polymerization-induced self-assembly (RAFT-PISA): current status and future directions. Chem Sci 13:4192–4224. https://doi.org/10.1039/d2sc00762b

Yang B, Chen Q, Liu X et al (2020) Effects of pest management practices on soil nematode abundance, diversity, metabolic footprint and community composition under paddy rice fields. Front Plant Sci 11. https://doi.org/10.3389/fpls.2020.00088

Chapter 10
Relevance of Nanopesticides and Nanofertilizers in Sustainable Agriculture

10.1 Introduction

In recent years, an increasing number of nanopesticides have been developed with the basic purpose of improving pesticide efficiencies and reducing their impact on the environment and human health. However, knowledge about their environmental interactions is still very limited. In a recent study, efforts were made to compare the soil fate and bioaccumulation of four model nanopesticides (namely, chlorpyrifos and tebuconazole loaded on polymeric and lipid nanocarriers) relative to the conventional formulations and pure active ingredients in containing earthworms and lettuce (*Lactuca sativa*) in two soils over a period of 4 months. These nanoformulations increased the soil half-life of the pesticides by twofold as a direct consequence of the slow release of the pesticide. Pesticide bioaccumulation in earthworms was often increased, probably as a result of their enhanced bioavailability. The pesticide bioaccumulation was also shown to affect plants, but with an inconsistent trend. Overall, the microcosmic results demonstrated that nanoformulations can significantly alter the fate and bioavailability of pesticides. However, generalisations are still difficult to make as the impacts depend on the type of nanocarrier, pesticide, soil used, and time scale. It could only be concluded that this study helped in judging the benefits versus risks of nanopesticides.

Over the past decade, new types of plant protecting nanopesticides have been developed using nanosized materials to possess novel technological, toxicological, or environmental features. Among their multiple goals, the emphasis has been placed especially on the following:

- improving their stability, solubility, stiffness, adhesion, and miscibility;
- increasing their effectiveness and specificity against pests; and
- decreasing risks via higher degradability and avoiding toxic solvents.

A. Nazeer et al., *Targeted Delivery of Nanopesticides and Nanofertilizers in Sustainable Agricultural Farming*, Nanotechnology in the Life Sciences,
https://doi.org/10.1007/978-3-031-41333-9_10

The organic nanocarriers are generally used in the form of nanocapsules or nanospheres of synthetic polymers, natural polysaccharides, or lipids loaded with an active ingredient. It is believed that the use of NFs could reduce the amount of pesticides entering the environment, decreasing their impact on the environment and human health. However, it is not yet clear whether NFs will result in such significant benefits relative to conventional products or whether they will add a new risk associated with NPs because of a lack of sufficient knowledge concerning their environmental fate and effects. It might be expected that the behavior and effects of NFs in the environment will be much more complex as compared to conventional pesticides. The nanoscale features will probably alter the environmental fate (sorption, degradation, mobility, and persistence), exposure (bioavailability and bioaccumulation), (eco)toxicity, and risk of an active ingredient. The colloidal properties of the nanocarrier and release kinetics of the active ingredient from the nanocarrier will have a significant influence. It is also quite likely that some targeted features that increase the effectiveness against pests can increase the environmental risks at the same time in terms of their increased persistence, bioavailability, enhanced mobility, and transport through biological membranes.

Over the last decade, an increasing number of studies have been directed toward developing these novel NFs, with the majority of them focusing on formulation preparation and very little on changing their effectiveness against pests. Only a few studies have looked into the effects of nanoformulation-related changes on environmental fate or the effects on non-target species. From the viewpoint of study design and the correct interpretation of the results, comparisons of NFs with conventional formulations or with pure active ingredients are lacking. Moreover, no study published so far has attempted to study the fate and effects of NFs under more realistic and complex conditions (Fojtová et al. 2019).

It was concluded from the above-mentioned study that pesticide nanoformulations affect their behaviour in soil and the uptake by earthworms and plants. The type and extent of changes are apparently dependent on the nanocarrier, compound, soil used, and application time. In several cases, the conventional formulations also change the fate and bioaccumulation of the studied compounds, which is in contrast to the assumption made by those assays about the environmental interactions of the pesticides. The NFs studied were able to increase the soil half-life of pesticides by up to four times and increase the residues in the soil by up to two times over the course of several weeks after application. Such increased persistence of the active ingredient (AI) and the consequent longer exposure to the pests might be one of the desired goals of the nanoformulations. However, it unfortunately appears that the NFs were able to alter the bioavailability and exposure of non-target organisms as well. A multiple increase in bioaccumulation in earthworms was often observed, probably because of the decreased sorption and facilitated bioavailability. In plants, several cases of bioaccumulation of NFs and AI in roots were found, but both an increase and decrease were observed, with the processes underlying them unknown. Overall, NFs showed a trend of having a greater impact than conventional forms, which is promising for applications but should be better controlled to ensure that only the benefits are enhanced and not the risks. In addition, it has been observed in the published literature that NFs have better pesticide activity related to their conventional and free forms, which could reduce the concentration of application and consequently the non-target effects. This might compensate for the feature of enhanced uptake of nanopesticides. Further studies are required to explore in detail whether the bioaccumulation changes induced by NFs are linked to shifts in (eco) toxicity and risks. It is very important to assess the fate and uptake of nanopesticides over a sufficiently longer time period. The conditions of the microcosms as a whole might make these observations more realistic and very valuable for capturing some of the complex interactions occurring in the field. This study is another proof of concept that NFs change the environmental properties of pesticides, along with the subsequent similar studies involving other pesticides and nanocarriers, contributing to a knowledge base to evaluate the benefits and risks of emerging nanopesticides and giving valuable feedback as suggested by several research teams.

With these kinds of inconclusive observations made in general, it is still pertinent to examine the future prospects of developing NFs with clear advantages exceeding the anticipated risks, as mentioned above.

10.2 Nanopesticide Technology

Extending the discussions further in the context of assessing the current situation and future prospects, it may be noted that a variety of compounds such as nanopesticides, nanoinsecticides, nanoemulsions, and NPs were developed using a nanoscale materials platform to deliver agrochemicals and various macromolecules needed for plant growth enhancement and resistance to stresses. Smart delivery of agrochemicals is thus targeted at increasing crop yield by optimizing water and nutrient conditions.

Despite the observations of mixed responses from various NFs, it can still be assessed in terms of global market growth at a significant CAGR of 14.6% from 2020 to 2027.

Globally increasing demand for food production and increasingly used various pesticides around the globe are some major factors to drive the market growth over the forecast period. The global nanopesticide market is segmented by product type, application, end-user, and region. On the basis of type, the market is classified into insecticides, herbicides, fungicides, and nematicides. In terms of revenue Research Corridor (2020), the insecticides segment held the largest market share (in excess of 41%) in 2019 (Bratovcic et al. 2021).

Pesticides are used in agriculture to improve crop yield and efficiency. Urea is the most widely used fertilizer for crop production and a source of nitrate, nitrite, and urease that cause eutrophication in the environment as contaminants in water. According to the global urea market report, the demand for urea was 187.8 million metric tons in 2020. Nanopesticides offer a way to both control the delivery of pesticides and achieve greater effects with a lower chemical dose. The first type of nanopesticides is foreseen in the form of active ingredient(s) that are either manufactured nanomaterials such as metal NPs of silver and copper or NPs of zinc oxide, copper oxide, manganese dioxide, silicon dioxide, and titanium dioxide. The synthesis and application of nanopesticides aim to improve efficacy and durability with fewer active ingredients present, while minimising or eliminating any associated potential hazards.

Out of many nanoemulsion formulations of herbicides and pesticides, the major potential advantages of pesticides were noted in terms of their strong adhesion to surfaces, high penetrability, and broad range of applicability in comparison to other methods. The process of encapsulation has frequently been deployed, as it facilitates entrapping one compound within another, producing particles with diameters ranging over a wide range of nm to mm. The main benefit of encapsulation is that it ensures the short-term availability of nutrients in the soil, preventing them from reacting with other soil components and causing undesirable effects. The nanoencapsulation process has been used in the case of both pesticides and fertilizers. Nanoencapsulation of pesticides, especially, takes care of its hydrophobic nature, which provides greater stability, improved dispersion in aqueous media, and enables a controlled release of the active compound, which increases its effectiveness in controlling pests at reduced doses of treatment in crops and human exposure. The encapsulation of nanopesticides is generally of natural origin, such as polysaccharides (e.g. alginate, carrageenan, xanthan gum, chitosan, and modified starch), and proteins (chicken egg albumin, zein, casein, α-lactalbumin, β-lactoglobulin, collagen, and gelatine). Coacervates of proteins and negatively charged polysaccharides have also been used for the nanoencapsulation of various substances. Clay nanotubes (halloysite) have been developed as carriers of pesticides for their low cost, extended release, and better contact with plants, and they might reduce the amount of pesticides by 70–80%, thus reducing the cost of pesticides and also the impact on water streams. Polylactide nanocomposites are characterized by high sorption capacity and enhanced conductivity with respect to the pristine polymer. These nanocomposites are applicable at the plant-soil-water interface to increase the ion transport and sorption of nutrients. Nanopesticide residues containing nanosized or nanoformulated agrochemicals may be present in consumed products. The slow release of the active components in pesticides through nanocapsules or nanoemulsions implies a greater persistence in the organism that might cause greater risks.

Hollow mesoporous silica hybrids in spherical shapes were used as nanocarriers of cyantraniliprole (CNAP) to fabricate an adhesive nanopesticide (CNAP-HMS-PDAAM). The experimental findings indicated that CNAP-HMS-PDAAM showed long-term control efficacies against *Cnaphalocrocis medinalis* (Guenee) and *Chilo suppressalis* (Walker), mainly due to its strong adhesive property on rice leaves with sustained release capabilities. Twenty-eight days after spraying, the efficacy of all four doses of CNAP-HMS-PDAAM was significantly better than that of Benevia (cyantraniliprole, 10% EOD). In addition, the nanocarriers were biocompatible, with no obvious influence on the growth of rice. Broad spectrum antimicrobial properties of copper-based (nano) pesticides in agro-ecosystems might have unintended consequences on non-target soil microbial communities. Studies also found out the impact of a commercial $Cu(OH)_2$-nanopesticide, over 90 days, at single and seasonal agricultural application doses, in the presence and absence of an edaphic organism (the isopod *Porcellionides pruinosus*), on microbial communities' function, structure, and abundance. The nanopesticide applications resulted in significant changes in both bacterial and fungal communities, particularly in the case of seasonal applications. At the functional level, the nanopesticide caused a significant increase in microbial C utilisation and a significant decrease in β-glucosidase activity. This study discovered that the $Cu(OH)^{-}_{2}$ nanopesticide has an effect on the soil microbial community, potentially affecting its ecological role. Nowadays, researchers are focusing more on the application of silica NMs in agriculture. Nanosilica-based fertilizers, pesticides, and herbicides could possibly improve the growth and yield of crops. Silica-NPs can be easily inserted into plants to modify the plant's catabolism and anabolism (Bratovcic et al. 2021).

10.3 Nanofertilizer Technology

Innovative nanofertilizer technology is required for the encapsulation of micro- or macronutrients in nanofertilizers, which can be encapsulated by NMs, coated with a thin protective film, or delivered as emulsions or NPs. The application of nanofertilizers allows the release of nutrients into the soil gradually and in a controlled manner, thus preventing eutrophication and pollution of water resources. Nanofertilizers not only improve crop yield and quality with higher nutrient use efficiency, but also reduce the cost of production and thus contribute towards agricultural sustainability.

In the context of developing nanofertilizer technology, the strategy to enter the market should be via the development of high-efficiency products for the major elements, including N, P, K, Ca, Mg, and S, which are applied in larger volumes and are currently less efficient and more environmentally polluting in nature. In case it is possible to apply less kg per hectare and fewer applications per season with the use of nanofertilizers, the cost of fertilization per unit of fruit yield will be significantly reduced. Compared with conventional fertilizers, the nanofertilizers lower the rate of nutrient release and, on the other hand, increase their stability by aggregation or adsorption, but without causing changes in chemical speciation due to pH-dependent phenomena, redox potential, and the availability of ligands in the soil. Examples include the following:

- Deploying nanorock phosphate that increases the bioavailability of P for plants since it prevents the fixation of P by Fe, Ca, or silicic acid, and
- Application of modified zeolites with a high internal surface for ion exchange for the regulation of N, P, and K in the plants

An increased growth rate (32%) and seed yield (20%) of soybeans compared to those treated with conventional fertilizer confirmed the advantages of phosphatic nanofertilizers. SiO_2 and TiO_2-NPs increased the activity of nitrate reductase in soybeans and intensified plant absorption capacity, making it more efficient in water and fertilizer applications. Furthermore, nanofertilizers may also be used for enhancing abiotic stress tolerances using microorganisms called nanobiofertilizers to provide great additional advantages.

Several studies confirmed the utility of nano-silicon dioxide application with biofertilizer was useful to increase the crop yield. A green chemistry-based synthesis of zinc, manganese, and iron oxide NPs using a microwave-assisted hydrothermal method with an average particle size ~20–60 nm was found useful as a foliar nanofertilizer on squash plants. The experimental observations showed that the spraying of manganese oxide NPs on the plants led to the best vegetative growth characteristics as well as the characteristics of the fruits, yield, and content of photosynthetic pigments. On the contrary, the content of organic matter, protein, lipids, and energy showed the highest value in squash fruits that have been sprayed with iron oxide nanoparticles. Iron and manganese NPs were biosynthesized through a simple and rapid method using bacterial supernatant containing the auxin complex (indole-3-acetic acid, IAA) for preparing the plant nanofertilizer. The synthesized NPs showed potential as micronutrient fertilizers for crop production. Bimetallic MnOx/FeOx-NPs from bacterial supernatant showed the best result on plant growth, especially in germination rates, root growth, and fresh weight in maize plantlets.

Foliar application of NPK nanofertilizers at a 50% rate in potato production increased production, profit, and quality compared to soil applications. The results showed that using lower rates of nanofertilizers in foliar applications was found to be an eco-friendly and economic alternative to recommended chemical fertilizers with a significant increase in potato productivity and quality. The potential effects of using two engineered NMs, namely nanochitosan (Cs) or carbon nanotubes (CNTs), as fertilizers on French bean plants, either alone or in combination with NPK, showed improved growth, yield, antioxidants in plants, and biochemical content of yielded seeds than seed priming. Foliar treatment, however, increased the duration before harvesting without reducing yield by 37.5% (80 days) as compared with control and seed priming treatment (110 days). Cs-NPs appeared to improve growth and yield parameters more than CNTs in foliar applications (Bratovcic et al. 2021).

10.4 Environmental Impacts of Nanopesticides and Nanofertilizers

The extensive release of NMs into the environment and the food chain might be anticipated to be risky to human health. Despite the fact that NPs' characteristic physicochemical properties may greatly influence their transformation and bioavailability during environmental spread, there is a lack of comprehensive data to establish assessments for the potential effects of pollution between toxicity and NPs in the environment. Although nanotechnology has numerous benefits to offer, the domain experts are still scared about the unsafe disposal of various NPs in massive amounts, like several hundred tons every year. The recent documents recommend that the presence of NPs could be detected in a number of controlled objects like atmospheric air, water objects, soils, hydrobionts, algae, fungi, tissues of land plants, and the tissues of animals. Until now, however, the potential negative impacts of NPs on human health have been rather speculative and not yet corroborated with concrete data. The current knowledge of ecotoxicological effects of nanopesticides is scarce, especially with regard to the fate and behaviour of such formulations in the environment. One of the key drivers for the application of nanotechnology is a reduction in the amount of pesticide needed to assure crop protection, which may be acquired in several ways, such as improved apparent solubility, controlled release, targeted delivery, enhanced bioavailability, and increased leaf adhesion. Studies reported similar toxicity, increased pesticidal toxicity, or similar toxicity at lower concentrations as compared to the conventional and nanopesticide formulations. Studies demonstrated that nanoformulations could possibly slow the release. Nanoformulations could enable better targeting of the pest. Nanoformulations could protect against a variety of degradation processes, including photolysis, hydrolysis, and soil degradation. The poor efficiency of the current fertilizers is a major issue to address. For instance, in the case of nitrogen fertilizer, up to 50% of the nitrogen applied is lost by volatilization and another 5–10% by leaching. Fertilizer loss has serious environmental consequences, such as eutrophication. Nanotechnology is used in plant nutrition with the aim of increasing the efficiency of current fertilizers, either by improving the delivery of poorly bioavailable elements (for example, phosphorus, zinc) and/or by limiting losses of mobile nutrients to the surrounding environment (for example, nitrate). The use of alternative growth promoters like TiO_2 or carbon nanotubes is also being explored.

Nanofertilizers can be put into three different categories: nanomaterials made of macronutrients, nanomaterials made of micronutrients, and nanomaterials acting as carriers for macronutrients. The first two categories comprise nanomaterials as nutrients themselves, whereas they are additives in the third category. Some scientists refer to the first two categories only as nanofertilizers, whereas the third category is termed "nutrient-loaded nanofertilizers" or "nanomaterial-enhanced fertilizers." While comparing the efficacies of nanofertilizers and non-nanofertilizers, assessments were made in terms of germination, plant growth, or crop yield observed in the application of nanofertilizers with corresponding conventional fertilizers. The resulting median efficacy gain of nanofertilizers over conventional fertilizers was 19%, 18%, and 29% for categories 1, 2, and 3, respectively. The category 2 nanofertilizers exhibited about 20% lower efficacy than their corresponding conventional analogues. This is probably due to the toxicity of several micronutrients at higher concentrations. The benefits of nanoagrochemicals are certainly improving the toxicity of pesticides to their target pest by up to 20% relative to conventional analogues. Also, published results show that nanofertilizers may increase crop production by some additional 20–30% compared with conventional fertilizers, so the advantages of crop protection and nutrition combined with reducing the use of agrochemicals by 20–30% could significantly mitigate environmental contamination.

The study to assess the relative environmental footprint of nanoagrochemicals is almost non-existent compared to conventional formulations. Given that there is general concern about the potential adverse impacts that nanoparticles may inflict in terms of both human and environmental health, more work would be required to evaluate the new risks and benefits that nanoagrochemicals would offer relative to existing products. As a result, current and future research must translate into the production of plant protection or nutrition substances, as well as the development of more sophisticated products that may aid in reducing the impact of modern agriculture on the environment and human health while contributing to global food security. Besides, evaluating nanoagrochemicals against products currently on the market will help avoid unjustified expectations associated with the use of nanotechnology in agriculture. Agriculture demands more innovations to meet the increasing demand for food while reducing its impact on the environment. It is vital that significant research efforts are invested in the development of novel products that are competitive and have the potential to make tomorrow's agriculture more sustainable. The application of nanotechnologies are expected hopefully to play an important role in reaching this goal.

From the previous studies, nanotechnology has been identified as a useful tool in modern agriculture to face emerging agricultural and environmental challenges principally related to the needs for increased productivity, sustainability, and security of agriculturally produced foods. The anticipated benefits of applied nanotechnology in agriculture may make it useful to implement agrochemical delivery systems in improving plant breeding, reducing the environmental impact of modern agriculture, and lowering input costs while increasing yield quality and quantity. While emerging nanotechnology concerns in agriculture include potential toxicological hazards and risks associated with NM-releases into the environment when used as nanofertilizers and nanopesticides, Despite the numerous advantages of using nanopesticides and nanofertilizers with controlled release of active compounds to improve plant nutrition and stress tolerance in order to achieve higher yields, not all nanopesticides and nanomaterials will be equally safe. Although the results of the research indicate all the benefits listed, however, the current level of knowledge about the ecotoxicological effects of nanopesticides is scarce, and it is necessary to examine the risks of the application of nanopesticides and nanofertilizers in more detail. In this context, further biotechnological advances are required for a correct and safe application of nanomaterials in agriculture (Bratovcic et al. 2021).

10.5 Chapter Summary

The accumulation of pesticides in plants is influenced by many factors, including temperature, humidity, plant species, plant growth, pesticide properties and formulation, and the mode of application. In one study, lettuce was exposed through the soil route, which explained the several times higher concentration of pesticide in the roots than in the leaves. Accumulation in the lipid-rich roots and only a small amount being translocated to the leaves is in agreement with the previous publications. BCF for TBZ was also greater in the roots than in the leaves, in agreement with the previous literature. However, considering the larger biomass of leaves relative to the roots, the results suggest an intensive translocation of TBZ from the roots to the edible parts of the lettuce. Interestingly, the ratio of TBZ mass in leaves to roots was several times higher for NF2 than for AI, which indicated a possible enhanced translocation of nanopesticides in plants. For a given soil, time, and formulation, the lettuce leaves always bioaccumulate less CLP than TBZ, and the lettuce roots bioaccumulate several times more CLP than TBZ. This illustrated the different bioavailability of both compounds in the soil for lettuce roots, as well as their different potentials to be translocated. However, the effect of the formulation was inconsistent for different times, soils, and compound variants.

Further Reading

Bratovcic A, Hikal WM, Said-Al Ahl HAH et al (2021) Nanopesticides and nanofertilizers and agricultural development: scopes, advances and applications. Open J Ecol 11:301–316. https://doi.org/10.4236/oje.2021.114022

Fojtová D, Vašíčková J, Grillo R et al (2019) Nanoformulations can significantly affect pesticide degradation and uptake by earthworms and plants. Environ Chem 16:470–481. https://doi.org/10.1071/EN19057

Chapter 11
Nanoenabled Agrotechnology: Current Status

11.1 Introduction

Nanotechnology has recently established significant potentials for nano-enabled applications and implementations in a variety of important sectors, including biomedicine, catalysis, energy generation and storage, exotic engineered nanomaterials, and extending its novel applications in agricultural science. Because of the quantum size effect, which changes their physicochemical properties in almost infinite ways, nanoparticles (NPs) with larger surface area to volume ratios have many more fruitful applications than their conventional bulk counterparts. The extraordinary functional properties of the nanoparticulate material species of inorganic, organic (polymeric), and biomolecular origins are being extended for their applications in sustainable agriculture due to their many nano-enabled product forms, e.g., nano-insecticides, nano-pesticides, and nano-fertilizers. Nanoparticulate materials have of late been considered an alternate for controlling plant pests such as insects, fungi, and weeds as well.

Although the basic understanding of nanoscience and technology is the same, it still requires numerous changes to meet the expected nano-enabled features and the actual specific needs. For instance, the extremely large surface-to-volume ratio of a nanoparticulate material sample not only imparts extended chemical, physical, and biological interactions during their applications but also numerous other features associated therewith depending upon the environment and the presence of atomic and molecular species that may be statically or dynamically changing in time. The simplest example of using surface reactivity has invariably been employed in the surface stabilization of the NPs, but other features like dissolution, aggregation, and conglomeration are also possible. This should also be noted because with every modification incorporated on the surface of the nanoparticulate samples, it establishes some characteristic properties associated therewith. Though the innumerable changes that can be incorporated appear to be very useful for designing

A. Nazeer et al., *Targeted Delivery of Nanopesticides and Nanofertilizers in Sustainable Agricultural Farming*, Nanotechnology in the Life Sciences,
https://doi.org/10.1007/978-3-031-41333-9_11

nanomaterials with controlled physicochemical properties, the overwhelming possibilities make it practically very difficult to fix nature, and thus it becomes mandatory to allow the molecular recognition to follow specific pathways that become more probable in order to achieve it.

When nanotechnology is considered for agricultural applications, the nature of interactions with the extremely complex composition of location-specific properties becomes even more complex to handle if not attempted with special care. For example, soils, being mixtures of organic, inorganic, and biopolymeric species varying vastly, have a very complicated but extremely active family of microorganisms already mentioned. To discover the biochemical interaction pathways of nanoparticulate samples with these microorganisms at the atomic and molecular levels, very elaborate experimental conditions must be designed. Although having some experimental data with whatever precautions one can think of is one thing, the situation in field experiments may be quite the opposite. Under all these extremely intermixed circumstances present in soil systems, the utility of the experimental observations is carefully selected to be least affected by other uncertainties. For example, the influence of different NPs on germination, plant growth, and antimicrobial features is carefully planned and conducted *in vitro* and *in vivo* before extending the observations at different levels of application.

Another characteristic feature of the agronanotechnology discipline is to try only those features that have been studied over a considerable period of time and collected from experimental investigations. Incremental changes observed due to exposure to different dosages of agrochemicals, comprising nanofertilizers and nanopesticides, are easier to track during experiments. Having compiled sufficient data from the ongoing experimental studies, one day it may be anticipated to apply machine learning, artificial intelligence, and other software-based *in silico* techniques to have a more meaningful picture.

Claiming to design a nanomaterial with predetermined features is still a far-fetched dream of scientists and technologists as it requires a far better understanding of the interactions taking place at atomic, molecular, and supramolecular complexes involving strong as well as weak chemical bonding along with Van der Waal's, hydrogen, and a variety of polar bonding studied mostly in isolation before applying them to practical situations.

The current efforts to grow certain crops in an artificially controlled environment are another way to isolate the extraneous influence of bioactive components that are invariably present. Hydroponic and aeroponic crops are good examples where the influence of micronutrients can be studied in isolation better than those attempted in normal soils, which are full of weather-related stress conditions. In order to identify the influence of stress-generating environmental conditions, researchers are currently studying them using genetic engineering approaches where detection of stress generating molecules, once detected, can be modified in time to apply mid-course correction rather than waiting for crop harvesting to study the harmful conditions.

The discipline of agro-nanobiotechnology is a vast canvas on which to attempt modifying various factors causing changes at atomic and molecular levels. Except for some basic observations established recently, attempts to understand various involved processes are not that simple. Investing in basic research and then extending it to commercially successful applications is a dangerous (not a simple) path to take. The projected requirements of food grains, fruits, edible products, and feed for animals involved in the farm sector are mind-boggling, and therefore adding any innovation to introduce even marginal changes may have far-reaching consequences.

The experience of developing synthetic fertilizers and pesticides, although it appeared very useful, produced its own side effects, which are well documented. As a precaution, it is better to evolve green nano-enabled products derived from natural products to minimise their negative impacts.

Antimicrobial properties of several kinds of NPs are currently being considered for smart and intelligent food packaging processes involving Ag-NPs. Apart from their antimicrobial properties, NPs of Si, Ag, Fe, Cu, Al, Zn, ZnO, TiO_2, CeO_2, Al_2O_3, and carbon nanotubes (CNTs) have also been found to have negative impacts on plant growth and development. Some of the salient features of nano-enabled products are examined in this chapter from different angles to highlight their emerging applications in the field-use of sustainable agriculture, along with their future perspectives and growing environmental concerns. One of the significantly important features of such engineered nanomaterial-based products is particularly related to their commercialized production at a very large scale for their wider applications in agriculture and allied sectors at affordable costs.

The impact of any growing technology slated for future applications is easy to assess by examining its R&D budget allocations. The United States National Nanotechnology Initiative (NNI) allocation of $27 billion in 2019; Nano-Mech, a leading NPs manufacturing company, receiving a $10 million allocation from Saudi Aramco Energy Ventures (SAEV); and the Massachusetts Institute of Technology receiving a $350 million grant to develop a state-of-the-art nanoscale centre called "MIT nano" reflect the anticipated impact of nanotechnology growth in the current times. The United Nations' ambitious goal of attaining sustainable development and achieving the zero hunger goal by 2030 reflects the critical requirement for a thorough variation in conventional agriculture practices by introducing the required changes by adopting the fast-emerging NT innovations.

Several significantly meaningful strategies have been developed by researchers over the last several decades to replace conventional agriculture practices, with the expectation that NT is one such growing methodology that is aimed at specifically modifying modern agriculture in sustainable ways in terms of enhancing crop productivity by taking care of the problems that limit the enhancements in plant nutrition, precision farming, water use efficiency, crop proteomics, and crop productivity. The unique physicochemical properties of NPs enable them to handle a wide range of applications in almost all industries, including agriculture, as well as numerous other field applications such as biomedical, chemical, optical, pharmaceutical, food, cosmetics, textile, etc. In addition, NPs have also been used in fortifications and food processing.

The commercial level entry of nano-enabled agro-products has already been recorded, especially in the formulations of nano-fertilizers for sustainable agriculture, growth stimulators, soil-improving agents, and nano-sensors, besides being used as smart and controlled delivery materials for reducing their overall costs during applications. However, the rapid applications of NPs are not rising without significant concerns for soil microbiota, accumulation in plants, and the threat to human health via the food chain. Any discussion in this context would be incomplete unless various aspects related to the safe applications of nano-enabled products for sustainable agriculture are examined, with an emphasis on the resulting environmental concerns.

11.2 Nano-Enabled Products: Field Applications

Although it is currently difficult to verify the proof of concepts that are mostly being tested at the laboratory scale, the practise of using NPs in agriculture still has a promising future. The synthesis of engineered NPs using their surface modifications for their agricultural applications has considerable potential to alter the perception of conventional agriculture by assisting in solving persisting challenges in agriculture using advanced materials and technologies. Most of these problems are usually related to pesticides, herbicides, fertilizers, and related formulations for their optimal uses. In plants, the distributions of NPs resulting from following some characteristic biochemical pathways are being explored with multiple advantages.

NPs have adequate capacity to protect plants due to their essential features, such as their extremely large surface-to-volume ratio, ultrasmall size, and optical properties arising from their characteristic energy band structure. The use of nano-encapsulated fertilizers and pesticides that are optimally dispensed using a method that distributes the content in a controlled manner is a significant accomplishment in this context. This method is also site-specific. For the protection of non-target organisms and other collateral repairs, this highly efficient method is beneficial and provides an effective defence mechanism when plants engage with pathogens by sending systemic compounds such as jasmonic acid, salicylic acid, benzothiadiazole, and others to the appropriate sites.

Some other advantageous features of NT have been noticed in attenuating natural pesticides and biopesticides, besides the controlled release of fertilizers supported by NPs, micronutrients, nano-sensors, and organic fertilizers. The nanotechnology products database (NPD) website, used to keep updated records on nano-enabled products, showed the total number of products to be 9245, manufactured by 64 countries via 2658 companies in commercial production. Various nano-enabled products and their applications have already been identified in different fields like agriculture, automotive, construction, cosmetics, printing, petroleum, the environment, textiles, electronics, sports, fitness, the environment, home appliances, renewable energy, food, medicine, and other nano-enabled products.

The nano-enabled products related to plant protection, fertilizers, soil improvement, plant breeding, and animal husbandry in agriculture are more than 230, manufactured by 75 companies in 26 countries all over the world. There are more than 906 nano-enabled products manufactured by 410 companies in 41 countries. A total of 892 nano-enabled products are in use in cosmetic industry. In printing, 664; in textiles, 814; in electronics, 894; and in pharmaceutical companies, 1121 nano-enabled products are in use. Thus, nano-enabled products have a wide range of applications in various industries.

11.2.1 Nano-Enabled Products in Crop Protection

Currently fungicides, pesticides, herbicides, and fertilizers are frequently used in soil in inappropriate amounts through foliar applications. Fungicides, pesticides, and herbicides are mixtures of chemical substrates that can be used to control, prevent, or destroy any pests, microbes, or unwanted herbs that infect crops and reduce crop production. Algicides, herbicides, fungicides, pesticides, and plant disinfectants are the main nano-enabled products considered for crop protection nowadays. More recently, nano-based stretch films made with polyethylene have been developed for hay packing to protect crops against UV rays and thunder and have been used in active food packaging systems to protect against *Escherichia coli* spoilage. About 40 nano-enabled plant-protection products are available in the commercial market in 12 countries and developed by 16 companies. Among all 40 products, 13 are widely used for crop protection and disease management.

11.2.1.1 Nano-Pesticides, Insecticides, and Herbicides

Pesticides kill unwanted, undesirable substances like fungi, algae, and bacteria and make plants resistant to them. The major difficulty experienced with pesticides is their solubility because they are not easily soluble in water and release very slowly after spraying, thus making them ineffective. Alternately, nanopesticides enclosed within a shell made with the support of NPs are currently being developed. These nanopesticides reduce soil runoff while also increasing solubility.

Liang et al. (2022) used intelligent response controlled release carriers to encapsulate active ingredients as a promising approach to increasing pesticide utilization efficiency while reducing environmental risk. Bonding pectin with metal-organic framework nanoparticles (FeMOF-NPs) loaded with pyraclostrobin (PYR@FeMOF-pectin-NPs) resulted in an intelligent redox and pectinase-based dual stimuli-responsive delivery system. The successful fabrication of PYR@FeMOF-pectin-NPs was proved by a series of physicochemical characterizations exhibiting a loading capacity of PYR@FeMOF-pectin-NPs for pyraclostrobin of approximately 20.6%. The pectin covering the surface of PYR@FeMOF-NPs could protect pyraclostrobin from photolysis. Different biological stimuli associated with

Magnaporthe oryzae could trigger releases of pyraclostrobin from the pesticide-loaded core-shell NPs, causing pathogen killings. The PYR@FeMOF-pectin-NP system outperformed pyraclostrobin suspension concentrate in terms of fungicidal activity and duration against *Magnaporthe oryzae*. In addition, the FeMOF-pectin nanocarriers showed no obvious phytotoxicity on the growth and could enhance the shoot and root lengths of rice plants. More importantly, PYR@FeMOF-pectin-NPs had an eight-fold reduction in acute toxicity to zebrafish than that of pyraclostrobin suspension concentrate. A dual-responsive FeMOF-pectin nanocarrier has great potential for achieving site-specific pesticide delivery while promoting plant growth.

Some insecticides that are not fully soluble in water must be dissolved in organic solvents, making them toxic and also increasing the cost of the pesticide. Using NPs could nevertheless enhance the solubility of insecticides besides reducing their toxicity. Modified chitosan and porous silica-based NPs have been effectively loaded into insecticides, and less water-soluble pesticides can be prepared. In *Spodoptera litura* ovarian cell lines, hydrophobic insecticides such as azadirachtin with improved chitosan NPs demonstrated a promising reduction in cell proliferation and the sustained release of drugs. The use of a solid–lipid mixture of NPs also solved the evaporation or volatilization issues correlated with the post-application of insecticide. Insecticides applied as a mixture with NPs were observed to be useful in slowing down the release of active molecules that could help to reduce the toxic impacts of insecticides.

The nanoparticulate formulation of insecticide emamectin benzoate (EB) was reported to be composed of DSPE-PEG$_{2000}$-NH$_2$ by the co-solvent method to determine its adverse impacts on the environment and to reinforce its dispersion, adhesion, and biocompatibility, resulting in encapsulation efficiency ($70.5 \pm 1.5\%$) of EB in DSPE-PEG$_{2000}$-NH$_2$ polymeric liposomes. The characterization techniques revealed that the DSPE-EB-NPs had a regular distribution, a spherical shape, and good leaf wettability, with a contact angle on corn leaves of 47.26° and a maximum retention that was higher than that of the reference product. DSPE-EB-NPs had strong adhesion to maize foliage and a good, sustained release property. The trials showed that the DSPE-EB-NPs had a strong control effect on *Spodoptera frugiperda* larvae, with an LC$_{50}$ of 0.046 mg/L against the third-instar *S. furgiperda* larvae after 48 h of treatment. All these results indicate that DSPE-EB-NPs could serve as insecticide carriers with lower environmental impact, sustained release property, and effective control of pests (Chen et al. 2022).

The mixtures of chitosan, silica, and polymer nanoparticles are the most explored and beneficial. The nano-fungicides have been noted effective against a variety of fungal species; however, they have rarely been tested on crops without any toxic effects. Insecticides have generic issues such as low water solubility, poor stability, and volatility, which can disrupt sustained release. These problems are solved by using fungicides based on NPs. Fungicides such as chitosan-lactide copolymers were loaded onto NPs in various concentrations, and their effectiveness was monitored as a growth inhibitor of *Colletotrichum gossypii*. An eco-friendly alternative to current synthetic fungicides is expected to play a major role in future plant disease management.

In vitro experiments for some repressive features of NPs such as Cu-NPs, CuO-NPs, Ag-NPs, and ZnO-NPs against economically relevant foliar and soil-borne plant diseases were conducted. The Cu-NPs were found to be the most effective in inhibiting mycelial development (EC50 values of 162 to 310 μg/mL), followed by ZnO-NPs (EC50 values of 235 to 848 μg/mL), and were practically insensitive to all fungal species, whereas Ag-NPs had a considerable inhibitory effect on *Botrytis cinerea*. Furthermore, when NPs and their bulk equivalents were evaluated for anti-fungal activity, ZnO-NPs were found to be more effective against all fungal species than $ZnSO_4$. Cu-NPs are more effective than those of $CuSO_4$ against all species, excluding *B. cinerea*, *A. alernata*, and *M. fructicola*. The lack of a link between NPs and their corresponding bulk metals, as well as the positive association between Cu-NPs and CuO-NPs toxicity, suggest that the mode of action of bulk and nano-sized metals may differ. There is considerable diversity although among the fungal species, the toxicity of all NPs rose significantly when applied to spores instead of fungal hyphae. This reveals that these compounds have a high potential for application as protective anti-fungal treatments.

Herbicides are used to kill weeds and unwanted grasses. They possess high levels of toxicity and a long half-life and are frequently unaffected by standard treatment. In general the weeds grown along with crops affect the growth and yield of the crops, and hence, traditional herbicides are applied to restrict the growth of undesired weeds. However, the application of herbicides might affect plant growth and development. In this context, nano-herbicides can be effectively used to solve such issues. The nano-herbicides are easily dissolved in soil particles, kill the weeds, and could be designed in a form that is less hard on crops.

Nano-enabled herbicides have the remarkable potentials to destroy weeds and improve crop yields. NPs mixed with certain herbicides, such as atrazine, ametrine, and triazine, are found to be 84% more effective as nano-enabled herbicides. Atrazine is quite often used herbicide to eliminate weeds and unwanted grasses. Once weeds are destroyed by herbicides, the nutrients available in the soil remain solely accessible for plants; consequently, plants develop resistance to diseases caused by nutrient deficiencies. Moreover, the use of modified NPs in combination with carboxymethyl cellulose could also improve pesticide breakdown efficiency.

Several nano-enabled materials are reported for weed control, also called nano-herbicides, and involve organic, inorganic, or hybrid materials. These nanoparticulate materials are able to control the release of metal ions and organic molecules in the agricultural field. Some studies have reported the ability of nanoherbicides to provide better weed management compared to non-nanoformulations. However, studies regarding the environmental risk assessment and mechanisms of action of nanoherbicides in plants are still at a very early stage of development. Nevertheless, they are essential to ensure a safe application for human health and the environment. A recent review discussed the current state of knowledge about nano-enabled herbicides, as well as some strategic design to fabricate nanostructures for weed control, resulting in their fate, behaviour, and effects in plants to achieve environmentally friendly and sustainable agriculture products (Forini et al. 2022).

Commonly used synthetic polymers like polyvinyl alcohol, propylene, polyethylene, and polycaprolactone used to develop nano-based formulations for agriculture applications might not quickly degrade, which brings some concerns regarding their environmental side effects when widely used. Using naturally derived polymers for developing nature-based nanoformulations has become more appealing with the emergence of green nanotechnology, which uses the principles of green chemistry for producing and applying these materials. Nature-based polymers are encouraged owing to their remarkable biocompatibility, biodegradability, low toxicity, and affordability, which consequently reduce environmental pollution. In addition, most of these polymers are soluble in water or less toxic organic solvents. In most cases, there is no formation of by-products and low energy consumption, ensuring sustainable consumption and production patterns.

In addition to maximizing agricultural outputs while minimizing inputs, nanotechnology has great potential for minimizing the environmental problems that can reduce agricultural production. It has been noted that deploying nanotechnology decreases the transport costs of the raw materials, makes the breakdown of raw materials like cellulose more efficient, and produces biofuels more efficiently, which results in the maximization of production per area and consequently can decrease the price of biofuels.

Priming is a popular approach, in this context, to produce plants with better and quicker responses to extreme weather events. Stress priming can increase plant tolerance to abiotic stress due to the induction of pre-established defence pathways that are unrelated to genetic modifications. Therefore, primed plants have a stress memory that leads to changes in the organism at the cellular and multi-omics levels, making them more responsive to recurrent stress within or across generations. Plants can be primed by stress elicitors or primary stress, boosting the defence mechanism linked to hormone pathways, antioxidant activity, redox signalling, biochemical modifications, and modulation of gene expression, which regulate the stress response. Furthermore, the elicitors used for priming are applied at low concentrations, which is advantageous over the costs of crop management.

The phenomenon of seed priming has been greatly affected by combining chemical elicitors with nanomaterials or using nanomaterials alone as elicitors. Owing to their unique properties, such as their small size, high surface energy and reactivity, and high surface-to-volume ratio, nanomaterials trigger several responses in plants that increase their growth, development, and productivity, even under adverse conditions. It is worth mentioning that each nanomaterial has different modes of action and consequently triggers different mechanisms for inducing stress tolerance in plants; however, the exact protective mechanism is unclear. In a nutshell, the seed nano-priming in plants has great potential for agricultural applications and endorses the SDG objectives (Campos et al. 2022).

Evolving more and more natural formulations hold promises to affect sustainable agriculture, as shown by the growing number of research reports available in the literature; however, the validation and commercialization of such products on the market are still in their infancy. Complete performance validation as well as regulatory agreement will be required for these products to be commercialized successfully. To be competitive in the market, nanoformulations must outperform their conventional counterparts by at least 20% in terms of efficacy, cost savings, or increased productivity. Although many nano-based formulations have a higher production cost when compared to conventional formulations due to the need for more sophisticated technologies and equipment, as well as specialized personnel, the financial return can be obtained with production gains. In a field study, Elmer and White (2016) demonstrated that copper oxide nanoparticles applied to the leaves of eggplants and tomatoes increased production while decreasing the disease symptom (*Fusarium* wilt). According to a profit margin analysis, a $50 per acre investment can increase production yields up to $10,000 per acre.

Another bottleneck of nano-based formulation application in agriculture is the safety criteria for non-target organisms, including humans and the environment. In contrast to nanomedicines, where exposure is individualized, agricultural applications of such products may affect a variety of target and non-target organisms. In this way, the safety studies of these formulations should be broadly evaluated. However, there is still no standardized and specific regulatory framework for the risk governance evaluation of nano-based formulations. To achieve standardized risk governance, regulatory communities worldwide must join efforts to reach a consensus on the definition of nanomaterials. In addition, developing reliable methods for identifying and characterizing nanomaterials in complex biological matrices, as well as monitoring their environmental fate and risks, has become increasingly necessary. Finally, advances in technologies are required to differentiate organic nanomaterials from organic particles existing in biological matrices.

Despite several advances made in nanotechnology for agricultural applications, the production of large volumes of nanoformulations remains an important challenge. In most laboratories, the volume of the preparation is a few litres. Furthermore, it is often challenging to produce large amounts while maintaining the same physicochemical properties. In addition, field and/or semi-field trials are rarely conducted, and if they are, the nanoformulations are only studied in one culture and are not reproduced over the years. Moreover, attention must be paid to the biodiversity of species cultivated in different parts of the world that can generate different physiological, biological, and molecular patterns and consequently cause varying responses to treatment with nanomaterials.

Studies should also focus on the nanomaterial concentrations required to exert biological effects. Evidently, the concentrations necessary to achieve the biological effect in seed treatments will be substantially lower than the concentrations necessary in the treatment of developed plants. Furthermore, dose studies should preferably be conducted under real-world conditions, where all factors that can positively or negatively affect the biological activity of nanomaterials are present. The effectiveness of the formulation is often one of the greatest difficulties in translating results from the laboratory to the field. One main reason for this is that the controlled environment of the laboratory where the tests are performed differs greatly from that in the real world.

Public acceptance is one of the main pillars of the success of a technology. Since the rise of transgenics and the fears and uncertainties generated by this technology in the population, the scientific community has introduced new technologies in a more cautious and less alarming mode. Therefore, education programs that spread the knowledge of nanotechnology to farmers who will use this approach and the general population that will benefit from its use are essential to reducing the uncertainties associated with this technology (Campos et al. 2022).

Santana et al. (2022) developed carbon nanomaterial-based target delivery for precise chemical delivery of CDs and SWCNTs and gene delivery platforms to chloroplasts involved in improving plant photosynthesis, assimilation of nutrients, and delivery of agrochemicals. Molecular recognition based coating of the nanomaterials with a rationally designed chloroplast targeting peptide was found to improve the delivery of CDs with molecular baskets (TP-β-CD) for delivery of agrochemicals and of plasmid DNA-coated SWCNT (TP-pATV1-SWCNT) from 47% to 70% and from 39% to 57% of chloroplasts in leaves, respectively. The percentage of dead cells in plants treated with TP-β-CD (20 mg/L) and TP-pATV1-SWCNT (2 mg/L) was 6% and 8%, respectively. The increase in H_2O_2 levels in leaves was associated with oxidative damage in whole plant cell DNA, a transient effect on chloroplast DNA, and a decrease in leaf chlorophyll content (−17%) and carbon assimilation rates at saturation light levels (−32%) with no impact on photosystem II quantum yield.

11.2.1.2 Nano-Fertilizer

The nutrient use efficiency (NUE) is a global concern because of poor efficiency of applied nutrients. Fertilizers are vital agricultural products that enhance crop production. In farming ecosystems, fertilizer contributes up to 35–40% of the productivity of any crop. Without the application of fertilizers to the crops, it becomes difficult to expect the required production. For this reason, a large amount of fertilizer is applied to crops in every country. Due to the fertilizer subsidies, the excessive use of fertilizer has increased the amount of nitrates in groundwater, which causes nitrate pollution. In recent past, the use efficiencies of N, P, and K fertilizers remained constant at 30–35%, 18–20%, and 35–40%, respectively, leaving a higher portion of added fertilizers to stay in the soil or enter the aquatic system, causing eutrophication.

Nanoenabled fertilizers could be used for regulating the release of nutrients depending on the requirements of crops in resolving the problem of fertilizer accumulation. However, now, only a few nano-fertilizer formulations are being synthesized in China, Germany, and the USA, and a few are being tested under laboratory conditions. The use of nano-fertilizers can increase the NUE and prevent environmental hazards. Recent research has also concentrated on carbon nanotube-based fertilizers. Carbon nanotubes have an extensive surface area and could regulate moisture under the constraints of irrigation or drought conditions. Currently, more than 102 nano-enabled fertilizer products are available in the commercial market in 17 countries and have been developed by 41 companies. Among all 102 products, five nano-products are widely used in crop production (Rajput et al. 2021).

11.3 Soil Health Improvement

Currently, NPs are known to be used in the soil system. After a detailed study of the soil, the researchers found a new class of clay particles called "Nanosols," which range in size from 1–100 nm and are more diverse than the other soil clay particles. *De-novo* synthesized NPs have novel properties and characteristics, i.e., a very high specific surface area and tiny surface charges that cause active reactions with other particles in the soil matrix. Due to their unique properties, these NPs are rather difficult to decompose rapidly and accumulate slowly in the soil. These nanoparticulate aggregates affect the soil properties. To reduce the NPs' toxic effects, it is better to combine NPs with clay as a colloidal solution. The constant use of urea, herbicides, and pesticides in the soil damages both the fertilizer capacity and soil health. When the concentrations of urea, herbicides, and pesticides in the soil exceed a certain level, they act like toxic pollutants and are ultimately harmful to both the environment and humans.

Recent studies indicated that the application of NPs improved the soil's enzymatic activities. Up to 200 mg/kg carbon NPs significantly improve the activities of urease, dehydrogenase, and phosphatase, and they also improve the storage capacity of soils for available nutrients such as ammonium N, nitrate N, and available P. Multi-walled carbon nanotubes showed a stimulative impact on these soil enzymes (urease, phosphatase, and dehydrogenase). NPs can play a decisive role in soil management. Nano-enabled products are normally available in the commercial market for soil management and can be found in three countries as developed by three companies. These companies' products (Descal, Humic 101, Coco: flower, Coco: grow, Hydro: flower, and Soilklik) do not add sufficient details regarding their product applications (Rajput et al. 2021).

11.4 Plant Uptake and Translocation of Nano-Enabled Products

The transpiration–cohesion–tension mechanism and root pressure make it possible to raise the water to the tops of plants through the vascular tissue system, primarily from xylem. Soil solutions with a higher water compared to root hair cells pass into cortical cells, through the endodermis and the pericycle, and then into the xylem vessels. This whole process of translocation depends on water potential. The exposure of NPs in the soil, likewise, first interacts with the plant cell wall and cell membrane of the root epidermis. Through a complex plant uptake mechanism, NPs enter the plant's vascular bundle (xylem) and translocate into the leaves. The absorption of NPs from the soil is also considered a primary step in the bioaccumulation of NPs in the plant system. Once NPs penetrate the plant cells and are transported into the plant tissue, there are two pathways for transportation known as the apoplast and the symplast.

The NPs are transported outside the plasma membrane through the extracellular spaces, cell walls of adjacent cells, and xylem vessels following an apoplastic pathway. Symplastic transportation of the NPs takes place between the cytoplasm of the adjacent cells through specialized structures called plasmodesmata and sieve plates. The NPs are translocated following apoplastic pathways for radial movement within the plant tissues that allows NPs to reach the vascular tissue systems that aid in further movement of the NPs through the xylem in various parts of the plant tissue. As the application of NPs in agriculture increases, it has become very important to know how they are transported and interact with plants' roots and organelles. Exact mechanism of the translocation pathways of NPs in plant cells could improve our understanding of the transportation of NPs via phloem or xylem. It may also aid in understanding their accumulations in plant organs that act as sinks, such as fruits and grains, so that in the future it is simple to detect NP accumulation and assess their associated toxicity in order to avoid the same by designing nano-enabled products to reduce nanoparticulate accumulations in plant cells. Consequently, it could be used to reduce their effects, such as by altering various biological processes (Rajput et al. 2021).

11.5 Concerns Associated with Nano-Enabled Products and Conclusions

The primary concern with the development of nano-enabled products is the potential concentration of NPs used in them, because higher concentrations of these NPs have been shown to have toxic effects at various levels. The continuous application of nano-enabled products, especially in agriculture, could increase their content in soil and crops as well. Most of the nano-enabled products examined in various studies do not reveal the concentration, particle size, name, or synthesis methods of the NPs used. Because nanoparticles can be transferred from root to leaf and leaf to root, they can easily enter the food chain and harm human health. They exhibit large-scale bioretention and accumulation, moreover, within living organisms, possibly beyond safe levels. Even a certain quantity of NPs could be harmful to human health. Higher levels of toxicity can cause growth retardation and inhibition. The toxicity of NPs is, however, specific to their shape, size, concentrations, base materials, and coatings. There is an extensive range of nano-enabled products in use that could be broken down into a different class of compound, such as metals, metal-oxides, carbon, and semi-conductor materials, and could generate new types of hazardous pollutants via their strong interactions. To address these issues, green nanotechnology is a multidisciplinary field that is creating clean, safe, and environmentally friendly alternatives to the products being currently used in various different industries. As their name implies, they are derived from different plant parts, microorganisms, amino acids, glucose, and similar other components. They suggest improving the commercial readiness of these technologies. Some current reviews have explored the use of nano-enabled products and discovered some interesting features based on recent findings. However, additional scientific evidence regarding these or other new nano-enabled products is still required. The studies should be directed towards determining the underlying mechanisms of nano-enabled products' interactions with the food chain, as well as their epigenetic consequences (Rajput et al. 2021).

Fellet et al. (2021) observed that a very profound transformation of the agricultural sector is required to become more sustainable and ensure universal access to healthy food. It would be essential to introduce radical technological innovations. Nanotechnologies have the potential to produce a significant boost to the improvement of the food system. In this context, in coming few years, a strong challenge will be faced regarding developing new and more efficient uses of nutrients in agriculture, with nutrient use efficiency (NUE) paramount in sustaining high crop productivity without depleting biodiversity or altering both the natural and agricultural systems. Nutrient leaching causes environmental pollution and water eutrophication, while nutrient excess favours pest and weed proliferation. Therefore, it will be mandatory to improve plant nutrition efficiency without affecting agricultural productivity or economic sustainability. A promising alternative is to introduce so-called "nanomaterial enhanced fertilizers" and "plant growth stimulators." Such innovation includes nanotechnological solutions that can improve nutrient delivery for a more finely tuned, accurate, and resource-saving distribution of nutrients. They also showed the latest advances in nanofertilizer research, mainly referring to nano-hydroxyapatite, silica nanoparticles, and chitosan-derived nanostructures.

11.6 Chapter Summary

The modifications in physicochemical properties and immense possibilities of biochemical interactions due to surface functionalization observed in nanostructured species (e.g., NPs) are currently being exploited in synthesizing nanofertilizers, nanopesticides, and nanoherbicides with commensurable advantages that can be put to use for their extensive applications in agriculture. Especially in the presence of a large variety of microorganisms in different types of soils used in agricultural farming, the situation becomes more involved. Moreover, farming being carried out as a part of a total ecosystem in which the environment plays an unavoidably significant role might some time lead to the apparent advantages turning into adverse effects on the plant as well as human health through various chains, including the food cycle. Some of the salient features to be considered in the case of sustainable agriculture are discussed in this chapter with supporting experimental observations to highlight the importance of nanotechnology in conjunction with biotechnology and agricultural practices. These considerations are finally amalgamated into a new discipline called agronanobiotechnology in this chapter.

Further Reading

Campos EVR, do Espirito Santo Pereira A, de Oliveira JL et al (2022) Nature-based nanocarrier system: an eco-friendly alternative for improving crop resilience to climate changes. Anthropol Sci 1:396–403. https://doi.org/10.1007/s44177-022-00029-x

Chen X, Qiu L, Liu Q et al (2022) Preparation of an environmentally friendly nano-insecticide through encapsulation in polymeric liposomes and its insecticidal activities against the fall armyworm, *Spodoptera frugiperda*. Insects. https://doi.org/10.3390/insects13070625

Elmer WH, White JC (2016) The use of metallic oxide nanoparticles to enhance growth of tomatoes and eggplants in disease infested soil or soilless medium. Environ Sci Nano 3:1072–1079. https://doi.org/10.1039/C6EN00146G

Fellet G, Pilotto L, Marchiol L et al (2021) Tools for nano-enabled agriculture: fertilizers based on calcium phosphate, silicon, and chitosan nanostructures. Agronomy. https://doi.org/10.3390/agronomy11061239

Forini MML, Pontes MS, Antunes DR et al (2022) Nano-enabled weed management in agriculture: from strategic design to enhanced herbicidal activity. Plant Nano Biol. https://doi.org/10.1016/j.plana.2022.100008

Liang Y, Wang S, Jia H et al (2022) Pectin functionalized metal-organic frameworks as dual-stimuli-responsive carriers to improve the pesticide targeting and reduce environmental risks. Colloids Surf B Biointerfaces. https://doi.org/10.1016/j.colsurfb.2022.112796

Rajput VD, Singh A, Minkina T et al (2021) Nano-enabled products: challenges and opportunities for sustainable agriculture. Plants (Basel). https://doi.org/10.3390/plants10122727

Santana I, Jeon SJ, Kim HI et al (2022) Targeted carbon nanostructures for chemical and gene delivery to plant chloroplasts. ACS Nano 16:12156–12173. https://doi.org/10.1021/acsnano.2c02714

Chapter 12
Nano-agro-biotechnology Prospects

Check for updates

12.1 Introduction

It is not yet easy to point out that adopting certain strategies would suffice to solve the multitude of complex problems being faced in the agriculture field, especially in terms of developing smart and intelligent agriculture practices based on optimal use of resources. The interactions of materials with the environment, including soil, aquatic, and atmospheric components in addition to the biodiversity of the biosphere, are found to be too complex to model any specific component of a bigger problem in isolation at all. The problem is manyfold more complex and highly interactive in nature than those in the case of human health care, where developing pharmaceutical solutions is rather better tractable under precisely known situations. Moreover, agriculture, being a component of the support system, has not been studied in the same detail as human physiology.

The pressing problems that are emerging fast due to global hunger and poverty with spiraling populations are putting enough pressure on creating a balance between rising food demands and an increasing population by making better use of the available resources, especially green chemistry, and nanotechnology. Despite having attempted the route of targeted deliveries using nanosized material entities, the problems are becoming more and more involved.

Although the problems facing the agriculture sector are daunting, solutions are foreseeable and will be achieved in due course with intensive efforts put into investigating the involved phenomena, both theoretically and experimentally, by involving researchers from all the related areas in a global attempt.

Adopting the 2030 Agenda for Sustainable Development, the UN committed to eradicating hunger, poverty, and achieving sustainable agriculture. Feeding humanity will require at least a 50% increase in the production of food and other agricultural products from 2012 levels by 2050. Growth in low- and middle-income countries is hastening a dietary transition towards higher consumption of meat,

A. Nazeer et al., *Targeted Delivery of Nanopesticides and Nanofertilizers in Sustainable Agricultural Farming*, Nanotechnology in the Life Sciences,
https://doi.org/10.1007/978-3-031-41333-9_12

fruits, and vegetables in comparison to cereals, resulting in commensurate shifts in output and adding pressure on finite natural resources. Low conversion efficiency of crop products to animal feed is another driver of the increased demand for crop production. Accordingly, agriculture must be more productive and diversified while coping with changes in climate and limited natural resources. A global challenge is related to producing more food with fewer resources, while preserving and enhancing the livelihoods of the farmers. Only emerging technologies are expected to help improve the current situation in agricultural practices. Despite long-standing discussions about using nanotechnology in agriculture, successful industrial applications remain limited in comparison to other fields. Despite the dropping of barriers like low public funding and the slow adoption of innovations by agricultural research organizations, a sudden increase in R&D has been witnessed in the past few years. Some of these emerging trends are discussed here in this chapter based on recent publications in this field, especially by Kah's group (Kah et al. 2019).

12.2 Overall Status

A variety of crops are produced for human consumption besides feeding livestock and the production of fuel (sugar cane), fibre (cotton), and other uses (rubber and tobacco), but only a small number of crops and countries dominate the global production. Although hunger and extreme poverty have decreased globally since the 1990s, approximately 800 million people are chronically hungry, and 2 billion suffer micronutrient deficiencies. Demand for both food and energy will continue to rise with the increasing global population and changing dietary patterns. The rate of agricultural yield increase has declined since the 1960s; for example, a 3% annual reduction in yield until the mid-1980s has declined to merely 1.5% in the case of wheat production. Moreover, food losses and waste have gone up to one-third of all food produced for human consumption, including 45% of all fruit and vegetables and approximately 30% of cereals. Even reducing the losses and wastes would require an increase in production, and this might be achieved through different strategies. In developing countries, large amounts of food are lost along the production and supply chains, mainly owing to a lack of infrastructure and insufficient equipment and technologies. In developed countries, additional waste occurs at the retail and consumer stages. Such waste in developed countries could be reduced by packaging and sensing technologies.

Even if losses and waste are minimised, productivity would still need to increase to match the growing demand. Average yields have increased since the 1970s with the extensive application of irrigation, plant nutrition, and protection, but still vary globally. Where biophysical and management conditions result in frequent stress, average yield might be even less than 20% of the yield potential, which is known in the case of maize in some regions of sub-Saharan Africa and Latin America. In many rain-fed cropping systems in developing countries, yields could be greatly improved with the existing technologies, but values remain low largely due to economic reasons, as the challenge of providing affordable tools becomes the main factor limiting yield and/or addressing specific deficiencies in the diet of local populations (for example, zinc deficiency in Africa and India). Many irrigated cropping systems, including maize in the United States, wheat in south Asia, northwest India, and Mexico, and rice in Japan and Korea, have yields that have plateaued at 80% of yield potential.

Yield gains in these regions are likely to be small in the future and might even decline due to climate change and the loss of arable soil. Raising average yields above 80% of yield potential seems possible with technologies that either substantially reduce the uncertainties in soil and climatic conditions involving sensor-based nutrient and water management or dynamically respond to changes in these conditions via targeted delivery systems. Such tools will also reduce costs and environmental impacts.

In spite of the increased yields due to better irrigation and the use of fertilizers and pesticides, these practices have their own limitations, resulting in an irrigation efficiency below 50%. Similarly, the conversion efficiencies for nitrogen and phosphorus fertilizers might be better than 10%. It has been observed that China, the world's largest consumer of nitrogen fertilizer, could lose more than 50% of the nitrogen due to volatilization and another 5–10% by leaching. Over the last 2–3 decades, the introduction of new hybrid crops and improved management have enabled improved nitrogen use efficiency in developed countries, but similar improvements in developing economies are still not feasible. Fertilizer runoff from fields could wreak havoc on the environment. Slower uptakes of nutrients by soils through fertilization with iron, copper, or zinc make it necessary to repeat the applications.

Extensive studies of foliar uptake and translocation of plant nutrients were conducted, but handling their complexity and variations proved difficult. The inefficient application and targeted delivery are still problems. It is known from experience that 10–75% of the pesticides employed do not reach the desired target. Similarly, non-target organism toxicity, in addition to increasing resistance, and their persistence and potential accumulation in environmental compartments, as observed in Europe, where the repeated use of copper fungicide on vineyards over a century has accumulated to an order of magnitude above local background levels. There would certainly be great benefits from the decreased environmental burden associated with crop production in the event that nanotechnology could increase the efficiency of irrigation, crop nutrition, and fertilization (Kah et al. 2019).

12.3 Nanoenabled Agricultural Practices: Constraints

The development of nano-enabled solutions for alleviating the problems faced in crop production envisages a number of constraints that are unlikely to change in the near future. Complex variability of conditions existing in the field like soil type, temperature, and water, as well as the existence of physiological barriers for the uptake and translocation of NPs in plants. Studies have demonstrated the uptake and translocation of NPs with diameters much larger than the commonly assumed size exclusion limit of the cell walls of 5–20 nm. Studies indicated that developing alternative modes of application of NPs for plant fertilization and protection holds great promise, including newer techniques such as seed coatings or direct injection into the phloem.

There are significant opportunities for hydroponic and/or vertical farming, particularly in terms of optimal land use, productivity, and food supply in urban areas. Agriculture, being a traditional industry with a narrow profit margin, will certainly need to invest substantially in developing innovative technologies to meet the growing food demand in a sustainable manner.

Increasing yield is not the farmers' concern; they generally aim for maximum profit from crop production. The costs of new technologies should be kept affordable even with the help of state subsidies to make them viable in a sustainable manner.

Difficulties faced by novel product developments that do not match the local legislative framework vary vastly from country to country. Creating some sort of legislative change would be necessary to foster innovative solutions to increase global production with minimum environmental impact.

Technological innovations in agriculture, being disruptive to existing practices, are seen as challenges by agricultural producers, and hence it is unrealistic to believe that nanotechnology will provide globally acceptable solutions. Rather, the most promising nano-enabled solutions should not only aim to solve well-defined problems but also contribute to the local legislative framework (Kah et al. 2019).

12.4 Current Status

In the context of having the desired material functionalities, nanotechnology would rather enable traditional approaches to offer better crop nutrition and protection. Combining the nano-enabled targeted delivery of the AI's of the existing pesticides and macro- and micronutrients with their delivery platforms would have a positive influence on the products currently in, or soon to be in, the marketplace. Accordingly, nanoscale platforms would be more appropriate to deliver new pesticides, biopesticides, and genetic materials for disease control (Kah et al. 2019).

Many of these approaches, as summarised below (Kah et al. 2019), would involve controlling the spatial and temporal delivery of the desired analytes in precision agriculture, in addition to providing new material functionalities:

- One approach uses NMs to improve plant nutrition as a defence against disease. Recently, NPs of Cu, Zn, Mg, and S have been shown to eliminate fungal, viral, and bacterial infections through enhanced nutrition, resulting in increased crop yield and nutrition. Experimental results are there to demonstrate enhancing the availability and transport within plants as a function of NP size.

- The elemental NPs might activate the barrier formations through resistance. Pathways have been identified with foliar-applied Ag/Ag-Si-NPs conferring pathogen resistance in tomato. A change in copper morphology from irregular to a nanosheet could reduce the effective dose by 10%. The coaddition of nanoscale elements could increase the efficiency of NPK use, as noted in increased N_2 loading into sorghum grain upon addition of ZnO-NPs, which acted as a nitrification inhibitor, reducing loss due to runoff or to the atmosphere.
- NMs are used in detecting plant pathogens, undesirable conditions, or stresses. Highly sensitive nanosensors were reported using boron-doped Si-NWs to detect proteins, calcium, and plant pathogens.
- The fluorescent silica-NPs conjugated with pathogen-specific antibodies were used as a nanoprobe to detect *Xanthomonas*, which cause bacterial spot on tomato. A fibre-optic based plasmon resonance immunosensor of Au-NPs was reported to detect orchid viruses. More recently, surface-enhanced Raman scattering based sensor was found capable of pathogen detection in food. In this context, it is noted that the implementation of sensor protocol must be robust in the environment and sensitive to the specific analytes.
- A molecule-specific detection of plant metabolites or exudates in the plant rhizosphere was developed to monitor crop health using a NP-based conductive system providing continuous real-time detection of stomatal aperture, implying the plant's water balance.
- Remote and *in situ* detections of pests/pathogens could reduce the uses of pesticides. It is worth noting that commercial deployment of sensors in agriculture has been limited, but given the large inefficiencies in agrochemical delivery, the topic of nano-enabled real-time sensing of growth conditions must be explored extensively.
- Similarly, nanoscale materials might also increase plant resilience against drought, salinity, and temperature, as noted in CeO_2 alleviating the salt stress in *Canola* in which the exposed plants had modified root apoplastic barriers and enhanced photosynthetic output enhancing the ROS scavenging in *Arabidopsis*.
- Similarly, Se-NPs enhanced antioxidant defence and improved sorghum yield under heat stress, although they resulted in toxicity at higher concentrations.
- Carbon-NMs could modulate plant responses, including increased germination and biomass, as well as greater salt stress responses, in a number of crops by altering the tomato fruit metabolome with due care and caution.
- The NP-co-exposure could reduce heavy metal and organic contaminants in crops; in most cases, this was related to ex-planta interactions between analytes.
- NMs improved soil structure and health, as evidenced by the use of nanoclays and zeolites to modulate water retention and release in soil.
- NMs were suggested to improve or alter enzymatic and/or microbial activity in beneficial ways. ZnO and CuO amendments could enhance overall soil enzymatic activity, causing negligible microbial community shifts.
- Iron oxide nanoparticles enriched the soil with iron without reducing phosphorus solubility for plants, as phosphorus reduction is a common issue with iron amendment.

- NMs are found helpful in crop biofortification, which will be more relevant as agriculture moves onto more marginal lands. The mechanism of action is related to the increased activity, availability, and dissolution of NMs. Se-NPs biofortify crops, but the nature of their experimental design confounds efficacy assessment.
- ZnO-NPs were found more effective than $ZnSO_4$ in promoting maize germination, vigour, and growth, resulting in 15% higher zinc levels in the plant. A similar biofortification of spinach with Se-NPs under field conditions was carried out.
- Pre-treating seeds with iron oxide nanoparticles improved germination and increased iron content. This could be a remedy for global Fe-deficiency once it is developed appropriately.
- TiO_2-NPs were found to reduce sugar and fat content in rice as atmospheric CO_2 levels increased. Similarly, an increase in CO_2 emissions will decrease protein, micronutrients, and vitamins in rice, causing negative health consequences. Given these projections, investigations into the sustainable use of NMs for crop biofortification are highly recommended.
- A variety of inorganic NMs, including metals, metal oxides, and clays; organic materials such as polymers; and novel materials such as GO and inactive viruses, have all been used as nanocarriers, though the NM itself may be the active ingredient.
- In the context of developing newer NMs, the synthesis of exciting species is expected to move ahead in the field of nanoenabled agriculture, especially by taking care of crop nutrition and protection.
- The development of smart, fast-responding, and advanced NMs would provide resources for meeting the specific demand of the crop specialist. For instance, a pH-sensitive, semi-hydrophobic coating with 15% COOH-group coverage of CuO-based fungicide incorporated into biopolymer was found to be degrading after 6 days of exposure to UV and moderate moisture. Perhaps the same could be combined with iron and zinc nanoparticles to promote crop biofortification.

Nanotechnology, holding great promise for addressing the major inefficiencies in agriculture and food production, requires vigorous research in all the related areas.

Although there are a growing number of reports on the preparation and properties of nano-enabled products for agriculture, the development and validation of such products for market deployment are mostly at the very early stages. Some nanoagrochemicals are already commercially available, like AZteroid FC and Bifender FC, Nano Grow 2.0, Nualgi Foliar Spray, and Dedalo Elite. Regulatory agencies indicate that they are receiving a growing number of product registration applications; thus, new nanoformulations are expected to reach the market over the coming years.

A patent search showed 3279 patents filed and 1254 granted in nano agrochemicals since 1990, in comparison to 16,757 filed and 6471 granted in pharmaceutical nano products. An increasing number of nanoenabled agriculture products are expected to enter the market after successful commercialization and comprehensive performance validation and regulatory compliance.

Besides meeting the need to demonstrate consistent performance or improvements in plant protection and/or crop yield, the nanoformulations must also meet safety criteria recommended by the regulatory agencies. The situation is more complex in the case of nanoagrochemicals, as they come into contact with several target and non-target organisms, like birds, bees, soil and aquatic organisms, and human beings. There is limited data on nanoenabled agrochemicals for field applications. There is a strong need for more field studies on the performance evaluations, fate, and impacts of nanoagrochemicals.

- The nanoformulations could enhance the efficacies by 20–30% above those of conventional agricultural products. Field use efficacy of a commercial nanoformulation for weed control was compared to the non-nano version of the herbicide (2,4-dichlorophenoxyacetic acid), resulting in a measurable increase (9–19%) in the yield of tomato, strawberry, spinach, and tea produce. A number of the nano-based treatments could achieve the desired goal of a 20% improvement in plant protection or crop yield, with the potential to offer functions that might open new markets where conventional technologies may not be available for comparison. For example, in the absence of sufficient conventional products, farmers can combine disease control agents and nutrients in simultaneous applications, in which the stabilized NMs enable complete pest control and nutrient applications. When applied to sugar beet crop, AZteroid FC (containing a nano polymer and the fungicide azoxystrobin), was found to be more effective than three azoxystrobin formulations due to stable AI in the high salt environment of the fertilizer. Similarly, applying starter resulted in increased potato and corn yields.
- A homoeopathic concept-based Nano-Gro product from Agro Nanotechnology Corp., using the addition of nanomolar concentrations of metal sulphates was found to enhance plant protection and crop yields.
- CuO-NPs sprayed on the leaves of young tomato and aubergine (eggplant) seedlings were effective in suppressing disease while significantly increasing crop yields.

The obstacles including lack of scale-up production processes, extensive field trials, and regulatory requirements are currently faced by the developers of nanoenabled agrochemicals. The difficulties associated with each step leading to market deployment could be decisive in discouraging small companies from pursuing it further.

Widespread adoption of innovative nanotechnologies in agricultural production will require a balanced analysis of the long-term environmental, social, and economic gains achievable by taking into account the sustainability and risks of handling nanotechnology.

Extensive studies of ENP safety evaluations should be encouraged in order to identify potential adverse effects with their use. Most developed countries require new agrochemicals to go through a rigorous human and environmental risk assessment before they can be sold. Current practices do not always capture nanoscale-specific behaviour, and adaptations may be required at different stages of risk assessment. There is currently no defined set of safety criteria that products must meet for approval by regulatory agencies with the breadth of nanoformulations and agricultural applications. Rather, nanoformulations for agricultural applications are generally considered on a case-by-case basis. For instance, the European Food Safety Authority (EFSA) recently issued guidance on the risk assessment of the application of nanotechnologies in the food and feed chains and recommended OECD documents for testing soil leaching and the toxicity of nanomaterials. The scope of current standard toxicity tests is insufficient to detect unintended consequences associated with the use of nanotechnology in agriculture. Advanced testing of chronic effects becomes important to ensure long-term safety. For instance, experiments looking at the possible trophic transfer of NPs or omics studies are essential not only to identify possible undesired effects but also to identify new strategies for the successful application of NMs for crop health (Kah et al. 2019).

Studies on the impact of NPs on microbiomes also allow a more comprehensive consideration of the soil, plant, or gut health status. Considering the very large amount of information and multiple endpoints often considered, it is important that the results be confirmed across species and environmental conditions before making generalizations. The use of agrochemicals always comes with a certain level of risk, and it is thus essential to compare the safety of new nanostrategies systematically with that of existing ones to critically assess the new risks and benefits. Risk evaluation currently focuses on the preparation and application of agrochemicals in the field. More comprehensive life-cycle analysis must be taken into consideration during production, transport, disposal, and the associated use of energy and water. This will also provide useful information that can help direct research and investment towards the most promising and beneficial strategies.

Finally, social aspects, including public perception and acceptance, should be carefully considered for the successful application of nanotechnology in crop production. Combining food with "nanotechnology" and "pesticides" or "fertilizers" has the potential to draw the attention of the public with concern. The risk of rejection in some geographical areas exists, partially fuelled by recollections associated with the partially successful introduction of genetically modified organisms in the 1990s. Public acceptance will necessarily grow once the benefits of the technology to society are demonstrated and as the need for advances in food production becomes more critical, highlighting the importance of pro-active and transparent communication to ensure understanding and acceptance (Kah et al. 2019).

12.5 Chapter Summary

Nanotechnology is certainly going to play an important role in the future of agriculture, including maximizing global food production and increasing sustainability, as noted from the above discussions. Inefficient practices existing today would certainly push for nano-enabled advances. However, finding solutions to the problems of inefficient use of fertilizers, pesticides, energy, and water has been a gigantic task due to the incredibly diverse and complex ecoagrosystem.

- Therefore, it is necessary to consider convergence-based nano-enabled solutions by combining the knowledge from agronomy, chemistry, plant physiology, material science, toxicology, and ecology, among others, including other areas like biotechnology, information and data sciences, artificial intelligence, and broad social sciences for their advantages.
- Full participation of agricultural producers right from the early process development would certainly be desirable to make the tailored solutions to meet their needs.
- Public outreach is necessary in promoting the technology acceptance by making them see beyond the common perception of nanosystems possessing greater risks compared to their non-nano equivalents.
- A comparison can be made between the nano-enabled strategies and those of current agrochemicals and good agronomic practises in terms of gains comprising economic, environmental, and/or food security aspects. A critical evaluation is important to ensure that efforts and investment are appropriately directed towards those applications that are competitive and address existing problems.

The assessment of new risks and benefits must evolve into an integrated approach that considers life-cycle, regulatory, and social factors. Developers of nano-enabled strategies should include safe-by-design and green chemistry methodologies so that the products not only meet the safety requirements for occupational and environmental exposure but also provide attractive alternative solutions.

Finally, it is the responsibility of the decision makers to find the right balance between the precautionary principle and promoting innovations for solving pressing problems faced by our society, especially when farmers do not have access to the innovations, technology, knowledge, and information needed to enhance productivity and incomes.

Further Reading

Kah M, Tufenkji N, White JC (2019) Nano-enabled strategies to enhance crop nutrition and protection. Nat Nanotechnol 14(6):532–540. https://doi.org/10.1038/s41565-019-0439-5